D1026411

DO CHOCOLATE LOVERS HAVE SWEETER BABIES?

The Surprising Science of Pregnancy

JENA PINCOTT

FREE PRESS

NEW YORK   LONDON   TORONTO   SYDNEY   NEW DELHI

FREE PRESS
A Division of Simon & Schuster, Inc.
1230 Avenue of the Americas
New York, NY 10020

First Free Press trade paperback edition October 2011

FREE PRESS and colophon are trademarks of Simon & Schuster, Inc.

For information about special discounts for bulk purchases,
please contact Simon & Schuster Special Sales at 1-866-506-1949 or
business@simonandschuster.com.

The Simon & Schuster Speakers Bureau can bring authors to your live event.
For more information or to book an event contact the Simon & Schuster Speakers
Bureau at 1-866-248-3049 or visit our website at www.simonspeakers.com.

Designed by Maura Fadden Rosental/Mspace

Manufactured in the United States of America

2   4   6   8   10   9   7   5   3   1

Library of Congress Cataloging-in-Publication Data
Pincott, J. (Jena).
Do chocolate lovers have sweeter babies?:
the surprising science of pregnancy / Jena Pincott.
p.   cm.
1. Pregnancy. 2. Childbirth. I. Title.
RG525.P496   2011
618.2—dc23   2011015005
ISBN 978-1-4391-8334-2
ISBN 978-1-4391-8335-9 (ebook)

To Una Joy,
without whom this book,
and many other things,
would not be possible

CONTENTS

7

EVE'S LEGACY, NIPPLE POWER, AND THE GOLDEN HOUR
Some Science of the Maternity Ward 158

8

MOMMY BRAIN, MOOD MILK, AND THE WEIRD HALF-LIFE OF CELLS THE BABY LEFT BEHIND
A Postpartumology 184

9

LESSONS FROM THE LAB
A Summary of Practical Tips 224

INTRODUCTION

O ne spring night at the end of my second trimester, I dreamed my fetus ran away. I woke up to find myself in an embryonic curl, the curtains blowing and moonlight streaming in. As the ceiling fan above circled lazily, I remembered this: my renegade baby, pinkish-pale, dashing out into a delicate and infinite landscape; the irresistibly charming little imp hooting and hollering and kicking the cancan; and me in hot pursuit, bumbling and breathing heavily.

"Every pregnant woman has body and self taken over by a chthonian force beyond her control," wrote the feminist writer Camille Paglia. "The so-called miracle of birth is nature getting her own way." I fear she's right. I feel my conscious, rational self isn't calling the shots anymore. I'm inhabited by urges and appetites that are not my own. I've asked myself: *I'm not in control, so who is?*

It's the fetus—at least in some ways, some of the time. The little ones manipulate us long before they're born. This helps explain some pregnancy weirdness, like how we might turn up our noses at delicacies we once loved (good-bye bitter greens, sayonara sardines) in favor of the sweet or simple foods of childhood. Some of us find ourselves desperately drawn to all that is safe and familiar. Why the eagerness to bond with our mothers, siblings, and girlfriends? When did we lose our edge, finding ourselves less open and even distrustful of the foreign and exotic? Why do we suddenly read fear, anxiety, and anger so often and easily on other people's faces?

Hormones are behind these developments. But who's behind the hormones? It's the unborn baby and her ambassador, the placenta. Changing Mom's behavior is a fetus's way of protecting her own interests: safety, food, and health.

I started researching this book before I got pregnant, wrote it as my baby girl grew inside me, and finished it several months after her birth. The nine chapters roughly span this time. A science writer and an expectant mom, I had been gripped by questions—some practical and others whimsical, some classic and others cutting edge—involving the science of pregnancy: Why are my dreams so vivid? Might what I eat now affect the baby's tastes in life? Is stress sharpening my baby's mind or dulling it? Could there be a hidden reason we conceived a girl and not a boy?

More and more questions bubbled up as the months passed. I ask whether pregnesia, or pregnancy amnesia, is for real. I explore whether there's a Daddy Gene and what shapes maternal instinct. I question what they say about newborns looking like their dads. I wonder if it's possible for my husband to nurse the baby. And by the way, are breast-fed babies really brainier? For answers, I've looked to hundreds of peer-reviewed studies and the researchers behind them. Evolutionary psychology, biology, social science, neuroscience, reproductive genetics, endocrinology—these scientific disciplines all touch on pregnancy, and I draw on them generously.

The science got weirder as I got rounder. I learned reasons that a partner's odor might turn us off and why to have sex with him anyway. Expectant fathers often find themselves plumping up and even puking, conditions triggered by their hormones. What triggers these hormonal tides? It's probably us—our behavior, or even our odor carrying chemical messages we produce only when pregnant.

A pregnant woman has more power than she realizes. Our ancestors believed that what we think, eat, and otherwise experience in pregnancy influences the baby in the womb—and there's increasing evidence it's true. Newborns are not blank slates. This is why scientists find it's worth investigating whether a woman's prenatal chocolate consumption might sweeten her baby's temperament, why kids are born favoring their mother's tastes and mother tongue, and how an expectant mom who keeps her competitive edge might sharpen her unborn baby's mind. The mother's sway over the fetus may begin at the moment of conception, or even beforehand, as I explore in the pages to come. Why else would bossy broads have more boys? And why do skinny chicks have more girls?

This book taps into the fascinating new field of epigenetics, the influence of environment on the behavior of genes. Recent studies in this emerging science reveal how blood sugar, stress and hormone levels, exposure to toxins, and even certain experiences alter how an unborn baby's genes behave without changing their underlying sequence. Epigenetics explains why babies of overeating moms, having developed in a womb larded with sugars and fats, may grow up to be hungrier, heavier, and more easily harried than they'd be otherwise. It gives insight into why it's foolish to frazzle a fetus, why smoking and starvation wreak havoc, and why our maternal instincts depend in part on how we moms were treated in babyhood. An epigenetic effect may even echo through the generations. If you ate a high-fat diet during pregnancy, for instance, your kids and grandkids might inherit a predisposition to diabetes. The dietary decisions your mother made when you were little more than an embryo could affect your baby too.

We're all immensely curious about what our babies will be like, and many of the questions in this book address prenatal prediction. If a fetus kicks a lot, will he be the active, fussy type after he's born? Might playing music or reading storybooks to our stomachs leave any impression? There are fortunes to be told by analyzing the baby's fingers in the ultrasound, predicting her birth season, and listening to the patterns of her heartbeat.

Looking at pregnancy through Darwin's eyes is intriguing—and even useful. I found instructive and thought-provoking theories on why we get morning sickness, why not to eat for two, and why a mother's bubble butt is good for her child's IQ. There are evolutionary reasons for why we can expect labor pains to intensify at night, how childbirth is both painful and forgettable, why we should seize the "golden hour," the first sixty minutes after birth, how breast milk is mood milk, and what we should expect from our new Mommy Brain. (Chapter 9 lists practical tips.)

This is not a comprehensive book of medical advice like *What to Expect When You're Expecting*. That's been done and done well, and there's no need for a rehash. While other pregnancy books are heavy on the how-to, this book focuses on the why—answering questions that our doctors won't touch. I write for the curious, inspired, open-minded, information-hungry mother-to-be (and the expectant dad too) who is seeking a deeper understanding of what's happening to her. I've presented the topics in an easy-to-browse question-and-answer format for the time-limited and commitment-phobic. It's curiosity driven: flip to the topics that interest you. I've tried to pack a mother lode of information into a format that you can read randomly and in small doses. I write with insomniac nights and waiting rooms in mind.

In the end, we learn not only about what's happening to our own bodies and minds during pregnancy, but also something about biology and human nature. It's all thanks to researchers who have peered into innumerable wombs, counted kicks and heartbeats, stretched headphones over huge bellies, and tracked men and women from the time they were embryos. They've crunched mountains of data and imagined what happened on the savannas where our ancestors evolved. Driven by curiosity and wonder, these scientists have analyzed the hormones of expectant parents, the genes of placentas, the fetus-friendly properties of semen, the flavors in amniotic fluid, the mind-control properties in tears, and the mood milk of laughing mamas. They've asked new mothers to press dirty diapers to their noses and tuck sweat collection pads in their bras. They've even entered maternity wards to find the truth about whether chocoholics have sweeter babies. It's all in the spirit of that one bold word that resonates with parents and children everywhere: *Why?*

DO CHOCOLATE
LOVERS
HAVE SWEETER
BABIES?

1

STRETCH MARKS, SHRUNKEN BRAINS, AND A MOST SURPRISING SMELL

Science Behind the Symptoms

Not long after our positive pregnancy test, my husband, Peter, and I visit the doctor for what should be a celebratory occasion: our first ultrasound. We're at ten weeks, nearly eleven on the obstetrical calendar, which begins the first day of the last menstrual period. By now I have taken on an ashen look. I'm shivering, my muscles ache. That morning I woke up from a light sleep with a heavy sweat.

"You're cold," my husband observes. I am perched at the edge of the exam table, stripped from the waist down and swaddled in two sheets. Peter throws his winter coat over me. I wrap his scarf around my neck and close my eyes. I'm a ball of woe. My problem is this: I've been bleeding and I no longer feel pregnant.

Until recently I had been feeling sleepy and sickish. I knew the hormone progesterone was partly responsible. Around the time my pregnancy test turned positive, progesterone would have been secreted from a crater in my ovary where that month's egg follicle had been. The crater, called the corpus luteum, is filled with yolk-yellow cholesterol, which is raw material to make hormones. Every month a corpus luteum waits for a signal from the egg that left on its womb-ward journey through the fallopian tubes. If it doesn't get a sign that a fertilized egg implanted, it self-destructs—which happens more often than not. Each cycle, the chances of pregnancy are only one in four among women in their early twenties, one in five in our late twenties, and down to one in ten or less by our mid-thirties and later. This month my egg, now an embryo, sent that rare, green light signal back to the mother ship: "I have landed!"

Once this happened, I started to feel pregnant. Homing in on the embryo's hormonal signal, the corpus luteum converted its supply of yellow cholesterol to progesterone and estrogen. Progesterone keeps the endometrial lining plump and intact. Damningly, its influence isn't limited to the womb. It also makes our breasts tender. It loosens our joints and ligaments, causing them to ache. It makes our stomachs sluggish, which means bloating, constipation, gassiness, and heart-burn. Progesterone is a natural sedative, yet we're peeing like race-horses (also blame the growing embryo, whose waste takes a toll on maternal kidneys). Progesterone makes eyes leaky too; we become moody. Someone sent me a sappy chain e-mail about a mother's love for her baby, now a middle-aged man. To my astonishment, I broke down into tears.

Odor aversions are another early pregnancy symptom. For me, the first culprit was my dear Siamese cat. Everything about the cat was agonizing—her food, her breath, the litter box. The smell of seafood similarly overpowered me. In week 5, at a French café that serves mus-sels, I slid back on the upholstered seat, my head throbbing and stom-ach roiling.

Odor and taste aversions are associated with the hormone human chorionic gonadotropin (hCG). The embryo produces the hormone, which rises throughout most of the first trimester. hCG stimulates the thyroid, which could lead to nausea and vomiting. The hormone has other roles too. What made the test line on my home pregnancy test turn positively pink? hCG. What signals the corpus luteum to start making progesterone? hCG. The hormone is how embryos communi-cate with their mothers, and it's in their best interest to turn up the vol-ume and make the signal as loud and clear as possible. Healthy embryos emit an ear-splitting amount of hCG.

As we wait for the doctor, I recount the facts: embryos that produce low levels of hCG frequently miscarry; embryos that produce high lev-els of hCG rarely miscarry.

The longer I sit naked in the cold, the more certain I am about my loss. I have no more hormonal symptoms of pregnancy—no more tenderness, no more sleepiness, no more queasiness, no more squeamishness. I imagine the embryo floating lifelessly, lost in space. I gulp back a sob. What happened?

Just as I can no longer stand the suspense, the doctor knocks on the door and steps in the room. Her eyes rest on me a moment as she adjusts the table. I give her a nervous little smile. Then I lean back as directed, my neck tendons cracking, and slip my naked feet in the stirrups. Peter grasps my hand. I hold my breath as the doctor inserts the ultrasound wand. In any other position than spread eagle I'd be openly fascinated by this experience. Ultrasound gives us a view of the womb by measuring the echoes of sound waves as they bounce off organs and tissues. But all that really interests me now is whether there's an embryo to see. It takes all my courage to look at the screen.

I see a bean-shaped smudge.

"A beauty," says the doctor. I am to understand that this is the embryo. Or *was* it the embryo? We need to hear blood flowing through a beating heart. A fetus's heartbeat may be detectable as early as twenty-two days past conception. The doctor presses another button.

BA-BOOM. BA-BOOM.

I let out a ragged little laugh and look up at Peter. Amplified and digitized, the heartbeat sounds as if it's coming from a healthy-sized alien. My husband beams with pride. I'm suddenly filled with a sort of swooning, breathy elation. I'm astonished.

BA-BOOM.

The truth is, symptoms come and symptoms go. Hormones rise and hormones fall. Bleeding may happen in first trimester, as it did for me—caused by implantation, a yeast infection, hormonal changes, tissue debris, or (my problem) a blood clot called a subchorionic hemorrhage that often dissolves over time. Some days I'll feel better than others. In the months to come, I'll experience changes in smell and taste, in perception and judgment, and even in memory. There are so many wishy-washy variables in how well we feel on any given day: diet, sleep, body chemistry, age, genes, and emotional well-being among them. Pregnancy, I realize, is about trusting our bodies more than our minds. Lesson to self: stop second-guessing self.

With a wobbly smile on my face, I hop off the exam table and get dressed. Just as I bend to put on my shoes, another surge overcomes me: nausea, in waves and ripples. I run to the bathroom, lean over the sink, and splash water on my face.

So this is how life begins, I realize: with both a boom and a whimper.

> It's quite clear that out of all of us, I'm
> certainly not the one in control. I am here to
> do your bidding, belly and babies. I am your
> humble servant.
>
> —Shannon Hale, author

IS THERE A PURPOSE TO MORNING SICKNESS?

Here are some of the paradoxes of pregnancy: The stomach is weakest when it needs to be strongest. Even as a fetus filches nutrients from our bodies, we tend to purge whatever we manage to consume. While you'd think we'd crave anything nutritious, it's often junk food that's irresistible. Why is it that veggies can make us queasy, yet devil's food cake goes down just fine?

None of this makes sense until we consider how even nice, wholesome foods can assault embryos. Meat may host bacteria and viruses. Cheese cultivates bacteria and fungi. Greens yield bacteria. Some vegetables contain compounds called phytochemicals that trigger an immune response, and some herbs consumed in excess may cause birth defects. You can tolerate tiny quantities of toxins if you have a mature and adequate immune system. But an embryo doesn't have one, so it needs to be protected. And the best way evolution can defend the embryo is to limit its mother's exposure to toxins by giving her nausea, vomiting, and food and odor aversions.

This is the embryonic protection theory, the most popular explanation for why so many women—75 percent or more—get nausea and vomiting in the first trimester. (*Morning sickness* is a misnomer; it's actually anytime-all-the-time-sickness.) A Harvard-educated maverick named Margie Profet came up with this theory in the 1980s, and it was controversial until two Cornell University researchers, Samuel Flaxman and his postdoc advisor, Paul Sherman, recently put it to the test. Analyzing more than seventy-nine thousand pregnancies around the world, they found a wellspring of support for the idea that queasiness, nausea, and vomiting have a purpose, which is to protect the embryo.

One bit of supporting evidence is that pregnancy sickness strikes when the embryo is most vulnerable: beginning around six weeks' gestation, and reaching the apex of agony between the ninth and fourteenth weeks when the bean officially becomes a fetus. Viruses, bacteria, and toxins could wipe the embryo out or cause defects and complications. This first-trimester window is also when the immune response temporarily weakens so as not to reject the embryo.

While we feel like we're falling apart, the embryo is gaining momentum. Its cells are dividing and differentiating, laying down the foundations of the brain, spinal cord, heart, ovaries or testes, larynx, and inner ear. There's a gorgeous medical term for this: *organogenesis*. By the time morning sickness has passed, around week 20, this process is complete. The embryo is now officially a fetus, and it has a human face.

Supporting the embryonic protection theory further, Flaxman and Sherman discovered that the foods that most often make pregnant women blanch—brussels sprouts, tomatoes, cauliflower, and kale, for example—contain naturally occurring toxins and phytochemicals. We have a decreased tolerance for anything bitter, which many veggies are in their natural state. Also topping the gag list are meat, fish, poultry, eggs, and caffeinated drinks. Nearly one in every three of us can't stomach them. Preventing infection is so important that it's probably why some pregnant women are dirt eaters—clay reduces exposure to bacteria, viruses, and toxins by lining and sealing off the stomach.

Think about foods a kid would like—sweet and starchy, white and tan. These you will probably tolerate and even crave. Least repulsive to a pregnant woman are foods high in carbs: grains and starches, fruit juices, cakes, and ice cream. Chocolate is one of the most commonly craved foods in pregnancy. And here's the finding that astonished me: if you live in a society that eats corn as a staple food, you're much less likely to get morning sickness. The corn kernel—the stuff of tortillas, cornbread, and popcorn—has remarkably low levels of phytochemicals, which means fewer natural toxins (explaining the number one craving in pregnancy: nachos). Meanwhile, Japan, a culinarily advanced country awash in seafood, has the highest rates of morning sickness in the world. When pregnant Americans travel there, they stick to plain noodles and McDonald's fries. Seriously, does the fetus, with its safe and simple tastes, hijack our tongues?

Odors also trigger nausea and vomiting. Although pregnant women are not more accurate at identifying smells, we have a stronger aversion to the ones we don't like. Frequent offenders are androstenone (in men's sweat), musk, lemon, leather, natural gas, orange, grape, cooking odors, cigarette smoke, perfume, and coffee.

Where I live, in New York City, reeks creep up from sewers and clobber me. I turn a corner, and one will slam me and send me reeling. I avoid: the fish market, the dog run, the toxic exhalations of the taco truck, dead rats, bacon-frying bodegas, the blue plumes behind buses, and the entrances of office buildings where smokers mope. Is it a coincidence that anosmiacs, people who have lost their sense of smell, often have symptom-free pregnancies?

Taste and odor sensitivity is usually attributed to elevated levels of estrogen, progesterone, and prolactin. These hormones coax the growth of new neurons in the brain involved in odor perception and affect taste receptors on the tongue. Estradiol may also make us headachy and queasy. Progesterone and prolactin slow digestion and make the stomach sluggish. Together they conspire to relax the sphincter of the esophagus, causing heartburn and other stomach complaints. Blood sugar levels, which tend to be lower in early pregnancy, are also culpable. The lower they are, the more delicate and nauseous we feel. (Incidentally, the worse the heartburn, the more likely it is that your baby will be born with a full head of hair, since the same hormones foster fetal hair growth.)

The tiny embryo itself may set off the strongest reaction—like the butterfly that triggers a tempest. The hormone most commonly associated with morning sickness is hCG. In early pregnancy, the embryo produces hCG (the placenta takes over later), and the hormone signal enters our bloodstream. It stimulates our thyroid gland, which can affect digestion and make us nauseous. The technicalities are fuzzy, so it's mostly guilt by association. As hCG steeply escalates in the first trimester, so does pregnancy sickness. When hCG levels settle down in the second trimester, so do the symptoms. HCG is an unborn baby's way of manipulating the mother to make his world a safer place.

Alas, we can't simply test hormone levels to predict how bad our nausea will be. Genes may also play a role: if your mother or sisters had pregnancy sickness, you're more likely to get hit. The more kids you have, the longer the pregnancy sickness. Oddly, pregnant women in their thirties or forties are less likely to have morning sickness than younger women. Their embryos may produce less hCG.

If it's a consolation to those worshipping porcelain gods, suffering has its virtues. Pregnant women who vomit have a lower risk of miscarriage than those who have nausea alone, as well as a lower risk of a baby with a congenital heart defect. One exploratory study found that women who had pregnancy sickness have up to a 30 percent lower risk of developing breast cancer later in life, thanks to a protective benefit of high hCG on breast tissue.

This isn't to say we should worry ourselves sick about not feeling sick. Most symptom-free women also carry babies to term, and those babies are just as healthy. My morning sickness is light and episodic even in the first months. Some days I feel a rise in my stomach, but many days I don't. I'm determined not to fret about whether I feel sick. I'm just going to ride the waves.

Do Girls Make Us Sicker?

When I tell people I don't have serious morning sickness, they tell me I'm carrying a boy. *Everyone* knows that boys don't make us as sick as do girls. Really?

Possibly. The girls-make-us-sicker theory is borne out in a few studies, including one that found that 56 percent of women rushed to the hospital for severe morning sickness later gave birth to girls. Those who were hospitalized for three days or more had an 80 percent greater chance of having girls. The explanation is that women bearing female fetuses have greater levels of hCG on average than those bearing males, and the more hCG, the harsher the symptoms. At every stage of pregnancy, the hormone levels are higher in women carrying girls, including the first three weeks, when hCG-based pregnancy tests are taken. Take note: If you're carrying a girl, you may get a positive test result earlier than if you're carrying a boy.

The trouble with using your hCG level to predict gender is that it's so unreliable you may as well flip a coin. Extremely high hCG can be caused by other factors, including twins or Down syndrome. Not all women with high hCG get sick (much less have twins or children with Down's), nor do all women carrying girls, and not all women carrying girls have higher-than-average hCG. Boys make us sick too. My own mother is a case in point. Although she was sick during both of her pregnancies, my brother made her much sicker than I did—so sick, in fact, that she may have had the condition that occurs in about 1 percent of pregnancies: hyperemesis gravidarum, or vomiting so severe that it leads to dehydration. So sick was she that her pregnant daughter feels queasy just hearing about it.

> There is no odor so bad as that which arises from goodness tainted.
>
> —Henry David Thoreau

WHY DOES DADDY-TO-BE STINK?

I once loved lox, but now it turns my stomach. My sensitivity to raw or smoked fish is understandable to people, and I get sympathy when I'm in a fishy-smelling place. But what my friends don't understand—what they think is almost too weird to talk about—is that I've developed a similar aversion to my husband's odor. Prepregnancy, I thought he smelled mellow and manly, like nickel and fire, corned beef and red pepper. But that was then and this is now, twelve weeks in. Why has he turned eggy and musky? Not every day, but all too often.

Oh, I know other couples have it worse. One pregnant woman told me she couldn't bear to sleep in the same bed as her partner. To protect her nose from sensory assault, she'd suck on industrial-strength lemon drops. Another said she had to keep her windows open whenever her husband was home. That winter was a brute.

If a partner's body odor has become a turn-off in pregnancy, it was probably a turn-on before. That's a good thing—a sign of compatibility on a biological level. If you once liked your partner's natural smell, it probably means he has immune system genes that complement your own. These genes are called the major histocompatibility complex (MHC), and they help detect and identify bacteria and viruses that invade the body. We advertise MHC genes in our body odor. Basically, these genes produce proteins that bind to odorants that ooze out of sweat glands in our armpits and genitals, where they mix with the bacteria on the skin. This mélange is as personal and distinctive as a fingerprint.

It's for healthy children that women prefer the smell of men with MHC genes different from their own. The mother's genes know the best way to beat some diseases, the father's genes know how to beat others, and their babies benefit by inheriting a mix of immune responses that can better protect them from disease. Children born to parents who are MHC similar often have a lower birth weight than other babies, are less healthy, and are less likely to have children of their own someday. Often sexually unsatisfied, women who marry men with similar MHC genes are more likely to cheat or fantasize about cheating. This is especially true during ovulation, when pregnancy is likeliest. They may take longer to conceive and also have a higher risk of preeclampsia, a disorder that may result in miscarriage.

Before I got pregnant, I liked my husband's smell, and that bodes well for our baby. It suggests that we don't share many of the same immune system genes. But we knew that. We once had our DNA analyzed in connection with a book I wrote. A TV producer taped us in our kitchen, swabbing our gums. Five days later, we received our results live on the show. Good news: of the MHC variants they tested for, we had none in common. We gazed at each other adoringly. My nose was validated.

So why does he stink now that he got me pregnant? One theory is that hormones mediate our attraction to other people's MHC odors, and pregnancy hormones reverse our usual preferences. Suddenly we're no longer drawn to people who smell different from ourselves. We like smell-alikes—people with body odors that resemble our own. Birth control pills mimic the hormonal state of pregnancy, and women on the Pill tend to prefer men who have MHC genes that are similar to their own. In both instances, pregnancy or on the Pill, we're not under the sway of sexy ovulation-related hormones that'd like to thrust us into the arms of MHC-dissimilar guys. Instead, we're under the spell of progesterone, a hormone associated with the desire to bond with others in a friends-and-family way. Between these various hormonal shifts, pregnant women may be more attracted to people who smell like kin and are more motivated to strengthen their relationships with them.

This is interesting. Could pregnancy subtly coax us to (temporarily) prefer our relatives over the baby's father? In our ancestral past, blood relations—parents, siblings, cousins, uncles, aunts, nieces and nephews—may have been more helpful than mates when it came to supporting a woman during pregnancy, childbirth, and child rearing. I wonder: is there a primal, unconscious, premonogamous part of us that seeks men with foreign genes (outsiders) to sire our babies and relatives (insiders) with similar genes to help raise them? It's just speculation, but this does happen in female mice. When pregnant, they prefer to nest (but not have sex) with others that have a similar MHC odor (kin) and shun those that do not.

Of course, we may go from loving to loathing our partner's odor for other reasons. Remember that our smell circuits have gone haywire. Some pregnant women become more sensitive to androstenone, the chemical in men's sweat and cologne that makes it smell musky, while others become less sensitive. What's your partner eating? A food that turns your stomach doesn't smell any better riding on a man's breath and body odor. Would he eat a blander diet? Would he hold off on the Old Spice?

The good news: I thought my husband smelled foulest in the first few weeks of my pregnancy. Now, toward the end of first trimester, the worst is over. The same goes for fish and rubber. Many women agree that the problem of a partner's odor fades as the weeks pass—perhaps we habituate to it—but for some, it lasts throughout the entire pregnancy. The problem will pass eventually. But don't hold your breath.

Love in the Time of Germophobia

Should a pregnant woman find herself attracted to people who never interested her before, or disgusted by those she once found desirable, she should take heart: it's probably just the hormones. Our tastes in men may change along with our culinary preferences. A lady once in the habit of eating samosas and tamales from food carts may find herself becoming a squeamish milquetoast, so to speak. She'll avoid anything potentially toxic or exotic.

During pregnancy, a subconscious fear of sickness or disease may influence decisions you make in your social life. In one fascinating study at the University of St. Andrews in Scotland, researchers asked more than one hundred women in their first or second trimester to rate male faces that varied in apparent health. Compared to a group of nonpregnant controls, expectant mothers gave significantly higher attractiveness ratings to men whose faces looked healthiest—having a ruddier color, smoother texture, and slightly stronger features. Pregnant women, it seems, value salubrity over character.

We have a good reason for this bias: we're more vulnerable to infection because pregnancy is a mildly immunosuppressed state. Just as we are repulsed by foods that could cause infection, so we may avoid faces that appear pale, pocked, or peaked. The researchers suggest that the hormone progesterone, related to bonding and kinship (and a lower sex drive), is behind these subconscious behaviors. Previous research has shown that when progesterone levels are high—when pregnant, on the Pill, or around our period—we tend to prefer faces and body odors of people who look safe and familiar and even somewhat feminine. It's the opposite of the type of guys we tend to favor at peak fertility.

This jibes with another study, which took place at Harvard University. Researchers found that pregnant women have a heightened bias against foreigners—finding them less likable, intelligent, and moral than insiders. The researchers speculate that we may have a biologically evolved disease-avoidance mechanism that associates outsiders with potential disease carriers. As with food and odors, disgust applied to people peaks in the first trimester. It's at this time that the exotic stranger we once desired to bed, we now just want to bathe.

DO WE SMELL PREGNANT?

It's week 13 now, and my husband and I are sitting in a cozy candlelit café. Over a meal of split pea soup and bread, the conversation inevitably turns to the pregnancy. It still feels unreal to me. No one has yet thought to ask me if I'm knocked up. I continue to have only a few low-grade symptoms: sore breasts, fatigue, insomnia, and occasional queasiness. I ask Peter if I look heavy yet, and he says no. For a moment I'm satisfied. Then I ask if anything else has changed.

"Your smell," he says matter-of-factly.

I lurch forward, causing napkin and bread crumbs to fall from my lap. "How? What do you mean?" My eyes are bulging.

"Hard to explain," my husband says. "It just has." He watches the elderly couple next to us with lively interest.

Et tu, sweetie? I've always valued Peter's candor, but this feels like an attack. It taps into my insecurities. I, too, think I smell stronger and gamier. But I thought that was just my skewed senses. As pregnant women become more sensitive to body odors, we might perceive our own in a new way. I didn't think a nonpregnant person, much less a man, even my husband, would notice. I like to think I live in an olfactory fun house; my odor perceptions are magnified and distorted, and no one else experiences them as I do.

This is true, but the research also suggests that our body odor objectively changes too. Blame the hormones again. We've become factories for progesterone and estradiol. By themselves, they don't necessarily have an aroma, but as they mix with the bacteria on our skin, they begin to produce body odor. A hormonal shift may also change the balance of bacteria, which changes the composition of body odor. (This is why the perfume we wore prepregnancy doesn't smell the same now.) The more we sweat, the stronger the smell. Pregnant women have a higher metabolism, so we sweat a lot.

Some studies suggest that heavy-duty progesterone, in particular, alters body odor in pregnancy. Women who take progesterone supplements often complain to their doctors that their sweat and urine smell stronger than before. In one study, male rats were significantly less

attracted to the odors of females that were given a single shot of progesterone, suggesting that the hormone inhibits the release of sex attractant odors (whereas an injection of estradiol increased sexual attraction).
If you're in touch with your body, you can detect a shift in your sweat,
urine, and vaginal fluid odors right before ovulation, when progesterone is low and estradiol is high, and after ovulation, when progesterone
surges, just as it does in pregnancy. (You might also taste different "down
there.") And yes, some men can, consciously or subconsciously, smell
the difference too, as found in studies in which male volunteers sniff
and rate the unlaundered T-shirts of women at different stages of their
cycles. No surprise: guys, without knowing why, tend to prefer the smells
of women when they are most fertile, not when progesterone is high.

That's not all. There's an even more surprising reason that your
body odor may be different: *you smell a bit like your unborn baby*. Even
a tiny fetus has an odor type—a distinct smell associated with the
immune's system's MHC genes. Your fetus's MHC odorants mix with
yours in your bloodstream, especially late in pregnancy, and create a
new fragrance of *le deux*. Your sweat smells different. Your kisses may
taste different.

Even your pee has a new stench because it contains the fetus's urine
too, as shown in an experiment by biologist Gary Beauchamp. Beauchamp and his colleagues at the Monell Chemical Senses Center paired
pregnant women in their third trimester with the rats in his lab. Known
for their keen noses, the rodents were trained to identify the distinct
urine odor print of expectant moms by touching a bar for the reward
of a drink. After the women gave birth, the researchers wanted to see if
the rats would still recognize maternal urine. The result was interesting: no, the rats showed no sign of familiarity. They also couldn't identify the odor of a mother's pee when mixed with that of a baby of the
same age who wasn't her own. Only when a woman's urine was mixed
with that of her infant, just as it was when the baby was in utero, were
the rats able to identify it.

"My smell. Is it *bad*?" I nervously ask my husband. He assures me
that it's fine—not bad, just different. As the saying goes, who is brave
enough to tell the lion that his breath smells?

But here's the part that makes me cringe. Because the baby inherits half her immune system (MHC) genes from her father, a pregnant woman's body odor will eventually contain *his* odorants too. Our sweat, urine, saliva, even our secretions down there—technically speaking, they're all tinged with odors related to the father of our child. Yet when a male has a sexual history with an expectant mom, he doesn't mind these smells. In fact, Beauchamp speculates that a man may be put off by the odor of a mate who is not carrying his fetus and that her odor could even trigger aggressive behavior on his part. Unfortunately, there are no human studies yet to test this theory,

The third reason we may smell pregnant to others, if even on a sub-conscious level, is that we sweat out a fragrant, pregnancy-specific mélange of volatile compounds that mix with bacteria on the skin and are released into the air. These chemicals have been detected in sweat under the arms and around the nipples, but only in the third trimester. An intriguing theory about these compounds is that they exist to help newborn babies identify their moms. Do our partners and others also detect their presence in our body odor? It's possible (see "Does Our Scent Affect Our Partner Subconsciously?" page 83).

Pets too seem to smell that we're pregnant. Among dogs and cats, females are often increasingly affectionate, although some become more nervous or aggressive. (My cat, Lavantina, is both protective and proprietary. She sits on my lap for hours. Her purr must be hardwired in my baby's memory circuits.) It's not much of a stretch to say that animals pick up on pregnancy-specific chemical compounds and hormones in our sweat. After all, many of these chemicals are also present in other species, from cats to clams, and trigger specific behaviors related to bonding and territorialism. Dogs are known to sniff out odors related to the immune system, which could be how they know we're pregnant (it's also how they detect tumors). Their sense of smell is 1 million times better than ours.

Compared to other mammals, human males have noses that are as effective as their ears, eyes, teeth, and tongues—that is, weak. Women are six times more likely than men to recognize their own body odor. Yes, the occasional male will notice a difference in his pregnant wife's smell. My husband did. But any guy who gripes about it would lack a more important kind of sensitivity.

> Everything grows rounder and wider and weirder.
>
> —Carrie Fisher

WHAT'S CHANGING OUR LOOKS?

Fourteen weeks pregnant, I accompany a friend to the doctor's office. This friend gave birth last year and wants her perfect body back. The doctor she is seeing is a plastic surgeon who works with lasers. She heard he is brilliant. He's the Merlin of bodily rejuvenation, a fixer-upper of butts and boobs, a magician who, with the wave of a laser wand or scalpel, tightens flabby arms, tucks tummies, and removes fat with such skill that he calls it not liposuction but liposculpture. On display in the waiting room are photos of sleek women displaying their curves.

I'm here to support my friend. But I'd also like to know what my body is in for and how I might prevent the worst of it. At the start of second trimester, the stretching has started. My breasts are getting so big they're itchy, and I worry about what will happen when they deflate. Last night at the gym, I noticed a great taut round of belly under my stretch pants. A heavy-set woman pointed at it and gave me a thumbs-up.

My friend, a workout fiend with porcelain skin and vibrant red hair, looks slim and refined. You wouldn't know she gave birth a year ago. "So, what do you need done?" I ask. She tells me she wants to look into the so-called mommy makeover: all-around lifting and tightening.

Oh, please, don't let this be me nine months from now. It's so easy to obsess about every bodily change. From the first weeks after conception, pregnancy advertises itself on the big screen: the skin. If you have freckles, they'll darken like bits of ember. If you have scars, they'll seem freshly seared. As the months pass, your nipples and aureola will slowly darken as if steeped in tea. The skin of the face, vaginal folds, gums, rectum, and scalp get duskier. A dark vertical line called the linea nigra may snake its way down your abdomen, from belly button to pubes. My linea is already tearing me apart.

Half of us will experience a blotchy brownish mottle known as melasma, or the "mask of pregnancy." Butterfly-like, it covers the cheeks and upper lip and can migrate to other regions of the face, neck, and torso. Melasma usually fades within a year of delivery. But parts pigmented before pregnancy, the nipples and vaginal lips especially, may forever remain a hue that ranges from lavender to blue-black with dark-skinned women.

All this is the effect of hormones on the hide. The skin, finely embroidered with receptors for estrogen and other hormones, is sensitive to their skin-darkening effects. From the second month onward, these hormones soar. Together, estrogen and progesterone stimulate pigment-producing cells called melanocytes that lurk in the bottom layer of the epidermis, the outer layer of skin. Once stirred, they surface and darken the skin. I feel like a slowly ripening fruit.

Other dark patches have a more exotic origin. Fetal cells can migrate from the bloodstream into the mother's skin tissue, resulting in rough papules and plaques on their bellies, thighs, arms, and butts. The more of your baby's DNA in your blood, the more likely you are to develop a skin reaction that manifests as spots and scales.

In addition to darkening, we may acquire striae gravidarum, stretch marks, on the skin of the abdomen, hips, butt, and breasts. The bad news: between 50 to 90 percent of us will get them. The risk of getting stretch marks has as much to do with heredity as the thirty or more pounds of weight gain. If your mother had them, you're more likely to be afflicted. The larger the baby, the more likely you'll get them. The good news: stretch marks become paler and less noticeable over time. Several delightful studies have found that expectant mothers over thirty years old are less likely to get striae gravidarum, presumably because older skin has less elasticity and stretches less.

But these marks are caused by more than stretching. The physical forces that expand the elastic fibers of the dermis are just part of the conspiracy. Hormones are in on this as well. There's the usual suspect, estrogen, as well as corticosteroids produced by the kidneys, and a ligament-softening hormone aptly called relaxin that acts on the fibers. By the third trimester, the layers of skin separate and bleed microscopically. As the dermis stretches, so does the epidermis. When the epidermis is stretched to the point of transparency, we see the purplish-red damage beneath. This will fade in the months after birth. Stretch marks heal as silver streaks and squiggles. They're several shades lighter than the skin around them because they lack pigment-producing cells.

What can we do to prevent stretch marks? After my friend leaves her consultation with a five-figure plan, I get a chance to meet the fabled Doctor Merlin. He gazes at me, or through me, as I pose the question. I tell him I'm still early in my pregnancy, but I can already feel the stretching, especially at night when I'm bloated. I'd like to know if there is any way I could prevent stretch marks now before they begin. Cocoa butter?

The doctor shakes his great head slowly. "Cocoa butter is folklore," says he. "Cocoa butter, shea butter, peanut butter, popcorn butter—it's all the same. Useless." But what does he know? He specializes in fixing problems, not preventing them.

Sadly, further research reveals that Doctor Merlin is right about the cocoa butter. No serious study has detected a difference in the severity of stretch marks between women who use cocoa butter and those who use a placebo. Optimistic or apocryphal, a couple of small studies suggest some benefit to vitamin E cream. In some of these trials, vitamin E is mixed with gotu kola, an anti-inflammatory herb, and in others it's combined with panthenol (vitamin B_5), elastin (a protein that works with collagen to make skin pliable), and hyaluronic acid (a sugar that lubricates the skin and joints).

Along with all this stretching, parts of the body are relaxing in preparation for the growing baby and the eventual birth. Progesterone, cortisol, estradiol, and relaxin soften joints and ligaments so that they're almost like rubber bands, which is why pregnant yogis tame their practice for fear of overextension and joint pain. Some are delighted that their feet spread like spatulas; we become more grounded, literally. There are women who go up a shoe size or two because the infrastructure of their feet has loosened. Female violinists and other musicians are startled when their precision wrists go floppy. Some of us will have difficulty walking when the ligaments of the pelvis stretch and sag like hammocks.

Also relaxed, and maybe too lax by the late second and third trimesters, are the blood vessels in the legs and rectum, resulting in varicose veins and hemorrhoids. Pressure from the uterus on the main artery, the inferior vena cava, causes those vessels to weaken, bag, and bulge. Exercise, elevate the legs, lie on the left, and avoid tight clothes. If your mother got varicose veins and hemorrhoids, you're more likely to get them too.

The same hormonal culprits show up again in other ordinary offenses: the cryptic mottling of the skin known as cutis marmota, which happens when exposed to the cold, and spider veins, which resemble tiny vampiric marks on the neck, throat, and face. If, around the fourth month, you spit out blood when brushing your teeth or biting into an apple, blame the effect of progesterone on the gums. This too shall pass after the birth.

It's not all terrible. In my first trimester and now into the second, some of the changes are flattering. I just reached a landmark birthday, so everyone says, "Wow, you look so young!" I'd normally respond with an eye roll, but there are days when I think so too. Some of it is my mood, of course—this was a long-awaited pregnancy, and I am ecstatic. But there is more to it. My face is plumper, and in a good way. I am not talking about fat, not yet anyhow. (Puffiness comes in the third trimester with increased salt and water retention.) The thin skin around my eyes and mouth has fleshed out, reducing fine wrinkles. My figure is fuller. Even my fingers, normally thin-skinned, seem more succulent. It's the dead of winter, and I don't have hangnails.

Estrogen has tissue-plumping magic more powerful than Doctor Merlin's. The face is especially sensitive to estrogen because it's rich in hormone receptors. Generally when estrogen levels surge, as they do around ovulation, many women think they look prettier. Our faces appear smoother, suppler, and more symmetrical. The same thing happens during pregnancy, at least to some lucky women. Yet I also unearthed an unsettling study that found that women with multiple pregnancies have more wrinkles later in life than their childless or single-child counterparts. Pregnancy may result in higher levels of hormone-binding globulins that trap free estrogens, and fewer free estrogens mean more wrinkles. Is this proven? No, but there are worse fates than smile lines.

Estrogen can clarify the skin, and sebum-secreting glands give us a waxen sheen. We radiate moon glow. To their astonishment, up to three-quarters of pregnant women report that they have less acne (others, like me, have more). As for the blush, is it maidenly modesty? No, pregnant women have increased blood volume and blood flow that cause tiny blood vessels right under the surface of the skin to dilate.

My bosom blooms. Breasts become bigger because the cells that make up the mammary ducts and lobules plump up under the charms of progesterone, prolactin, and lactogen, a hormone produced by the placenta. Women who are carrying girls grow larger breasts than those carrying boys. (If the baby is a girl, her breasts too start to grow as early as the fourth week.) My nipples are larger. On the areolae are tiny raised bumps, like the skin of a lychee. These are Montgomery's tubercles, and they secrete a protective lubricating antibacterial substance. They're like Venus flytraps. Look but don't you dare touch, I warn my husband. Fertilized by estrogen and progesterone, they swell, but the same hormones make them tender. The pain is the worst in the first trimester. After birth and lactation, breast ducts shrink as if bitten by frost. They batten down.

Estrogen and other pregnancy hormones are fertilizer for hair too. By the end of the first trimester, my tresses, while not exactly Rapunzel's, are fuller. Scalp skin is studded with hair follicles, each one lined with a hormone receptor that decides how long the strand will stay in place. There's the anagen phase, when hair grows, and the telogen phase, when the follicle is dormant and hair falls out. Estrogen extends the anagen phase, and hair stops falling out. Not only does hair grow more and shed less during pregnancy, but each strand is thicker. For us fine-haired women, a long anagen run makes us feel like fertility goddesses.

Estrogen plummets right after delivery, and within a month a new mother's lush locks start to thin out. The shedding goes on for up to a year after delivery, prompting many women to get the "Mommy cut," a hairdo that falls somewhere between the chin and shoulder. It's more practical, we may claim sheepishly, but the shearing conceals the shedding. Locks will likely thicken by the time the baby is fifteen months old, although some women never regain their prepregnancy fullness. Though by that time we're used to our mommy cut anyway.

Most pregnancy symptoms—the shedding, the stretching and the swelling, the bulging and the bleeding—reverse themselves months after delivery.

But, of course, not all.

Let's face it, Doctor Merlin will always have work. As I leave his office, I catch a last glimpse of him surrounded by his entourage of glamorous young nurses. Beauty standards are impossibly high. Why do we pathologize and stigmatize the postpartum body? I worked hard to have this baby, and when it's all over, I have a feeling I'm going to show a few zigs and zags no matter how much vitamin E I slather on. It's not only our bodies that need to adjust and accommodate. Our minds do too.

> Of all the things I've lost,
> I miss my mind the most.
>
> —Mark Twain

IS THE FETUS TINKERING WITH OUR BRAIN?

As the skin stretches, the mind shrinks. There's no more room in there for the names of people I just met, or for birthdays and anniversaries and directions. I can't remember the brand of my favorite bittersweet

chocolate or the item I'm looking for in the cleaning aisle of the grocery store. Everything seems slippery. Yesterday a woman asked me to
sign a book for an Erica.

"K or C?" I ask.

"C."

Fiddling with the pen, I noticed that my nails were growing faster
and longer than usual. The middle nail threw off my grip.

"Did you say 'k'?" I asked.

"No, Erica with a 'c.'"

I once babysat for an Erika, an eight-year-old blonde brat who
whacked off the heads of Barbie dolls. Teenage girls are worse. I
wouldn't have wanted to be the mother of my teenage self. I dread the
idea of fighting with my own kid someday.

I honestly can't remember if I ever signed Erica's book.

Between 50 and 80 percent of pregnant women feel forgetful and
absent-minded, according to the American Psychological Association.
It's called pregnesia, pregnancy amnesia. Naysayers claim pregnesia is
all in our minds, to which we the afflicted retort, "Yeah, and what
isn't?"

Over a dozen studies address the mental performance of pregnant
women. In one that took place in the Netherlands, expectant women
were given a battery of cognitive tests during their second and third
trimesters, and their results were compared with those of nonpregnant
women. Some of the tests measured behavioral planning (crossing out
items in the right order), information processing (matching numbers
with letters), and word learning (free recall of a list of words). Although
all the women sailed through the information processing tests at
approximately the same speed, pregnant test takers scored significantly worse on tasks that required memory encoding (storage) and
retrieval. Studies that required women to remember lists of words, forward or backward, or paragraphs or stems of words, found that their
word recall was significantly impaired. Many of the other studies came
to a similar conclusion: pregnancy impairs both immediate and
delayed recall.

This means we may forget our future intentions, like taking our prenatal vitamins or mailing a letter. We may have more difficulty remembering street addresses, passages of poems, phone numbers, and the names of people standing before us.

One small Canadian study found that women carrying female fetuses got especially low scores on tests of short-term memory, recall, and spatial relations. Their boy-carrying peers consistently outperformed them throughout the entire pregnancy from ten weeks to several months after birth. The researchers were struck by the difference. Why would girls mess more with their mothers' minds? Is it the estrogen, which is higher in women with female fetuses? Another theory is that women carrying girls often have higher levels of hCG, which may worsen morning sickness, and sick people don't perform well on tests.

Neither do zombies. Pregnant women have fewer cycles of restorative sleep during the night than usual, and what we get is frequently interrupted. How can we expect to be sharp without rest? I hardly remember what it was like to doze more than four hours without a sleep-shattering ache or bathroom break.

Brain scan studies give me more reason to lose sleep. Over the nine months of pregnancy, a woman's brain shrinks 4 to 6 percent, being puniest around her due date. The shrinkage doesn't mean we get dumber. We're not necessarily losing nerve cells; we're losing volume. Although scientists are unsure if the shrinkage is a direct cause of the foggy-mindedness, it may well be related.

Blame the baby. The little brain is stealing from the big brain, according to one theory. The greedy fetus and its ambassador, the placenta, commandeer phalanxes of brain-building fats to the baby's side of the border. Fetuses need to fortify themselves with omega-3 essential fatty acids for their own brain development. Coveted and pilfered, these omega-3s keep cell membranes fluid and affect every part of brain function. "Take more fat from my thighs!" I insist, but this is no democracy. In pregnancy, there is no Fourth Amendment; my bodily possessions are seized without recourse. I need these fatty acids, but so does my baby.

Surprisingly, one area of the brain that grows, not shrinks, in pregnancy is the hippocampus, the memory center. Animal studies have found that pregnancy triggers the development of new dendritic spines, which are budding growths that help transmit information signals between neurons. Some researchers argue that this new growth is a cause of pregnesia. You'd think growth would be good, but it may also mean chaos leading to impaired function—it's a jungle in there. The placenta's estradiol and progesterone are likely fertilizers for the new growth in the hippocampus.

A more radical theory is that the frenzy of new activity is triggered by the presence of something that boggles the mind: stem cells from the fetus that have found their way to the mother's brain (see "What Do Fetuses Leave Behind?" page 220). The lush new growth in the hippocampus may prep our brains for motherhood.

Yes, the fetus is taking over your brain, but it's a little like the joke when a man puts a gun to his own head and says, "I have a hostage." You and your baby are in this together.

It's not all so bad. Some of the cognitive changes are welcome. Take the effects of the hormone cortisol, for instance. Cortisol levels soar through pregnancy because the fetus needs it like a plant needs fertilizer. The hormone shepherds blood flow across the placenta and also helps the baby's brain and organs to develop.

While cortisol has a bad-boy reputation as a stress hormone, it's also a "pay attention" hormone. Although it's associated with memory loss, cortisol influences us to zero in on what's important to the fetus: environment, safety, food, and health. So we refocus, reprioritize. For me, hand-washing to prevent illness and hand-wringing about moving into a new home before the baby is born trump remembering the perfect bon mot or a phone number. Seven months pregnant, a lawyer friend of mine displayed unusual calm when she was hit with the double whammy of a pink slip at work and the theft of her car. To my astonishment, this workaholic shrugged her shoulders and stroked her belly, saying, "Ah, well, guess I'll be a stay-at-home mom for a while, and we won't need a second car." It's baby first.

In a way, the fetus is making us calmer and less fearful. By the end of the second trimester, once a threshold for cortisol levels has been reached, we often don't absorb subsequent stress signals much better than drenched sponges hold water. A dampened stress response protects newborns from toxic levels of cortisol. Pregnant rats are found to be much less anxious than virgins when confronted with challenges such as being dunked in a vat of water and forced to swim, and they're less likely to freeze up when thrust into an unknown space. A calmer, less fearful animal is more likely to leave the nest and forage for food to sustain her pregnancy. A more tranquil (and more forgetful) mother-to-be protects her baby from the dangers of excess stress (see "How Fetuses Calm Us," page 141).

Intriguingly, the fetus may also reduce our stress by encouraging us to get along better with others. Think about it. If you're about to enter this world, would you want your mother to be socially connected or a total outcast? A baby's chances of survival are higher if his mom has social support. Incidentally, the high progesterone levels in pregnancy are associated with a stronger desire for kinship and bonding. The more we bond, the more progesterone, and the more progesterone, the more we desire to bond. Perhaps this explains why I'm suddenly spending so much time on the phone with my mother and chatting up ladies about their nurseries and episiotomies. A close friend told me I now seem more grounded and approachable. I'm more in touch with others.

A recent British study found that by the third trimester, pregnant women have a vastly better ability to decode emotions in other people's faces—especially anger, panic, worry, and nervousness—than they did in the first trimester. According to the researchers, this may be an evolutionary adaptation to make vulnerable moms-to-be more watchful and alert to signs of threat, aggression, and contagion. We're better at recognizing faces, especially males, and telling who might help us and who might hurt us. I swear I can also smell emotions on people too—especially anxiety or embarrassment (sweat contains chemical signals for these hormones). Credit the pregnancy hormones (estrogen, progesterone, and cortisol) that act on the amygdala (emotional memory) and olfactory (odor-sensing) regions in the brain.

And who's the mastermind behind the mind reader? Again, it's the fetus, whose placenta is revving up the production of these hormones. Babies with calm, connected, intuitive, and vigilant mothers have an edge when it comes to getting what they need: food, health, and protection.

The lesson here is to get used to the fetus tinkering with your brain. It's evolution's way of making you a mother. Look at it this way: you're not losing your mind; you're gaining another.

> If nothing ever changed,
> there'd be no butterflies.
>
> —Anonymous

WHY ARE OUR DREAMS MORE VIVID?

If you expect sleep to be an escape from the weirdness of your waking hours, you'll be disappointed. Pregnancy dreams are notoriously strange and unforgettable. I'll give you a sample of the zinger I had last night. A traveler of sorts, I'm driving a rental car through a wooded landscape in a luminous dusk. The trees are skinny, white, and bare. The car has very little horsepower, and I barely make it over the hills and around the bends. Like a mirage, an open-air deli appears in a clearing. I stop despite the red-and-white roadside warnings. At the cash register, near the pickles and candies, a fleshless blond man with a blood spot on his gums tells me I am in Rhea. With horror, I recall that in Rhea, pregnant women are stuffed with adderberry and Rhime. Rhean souls are hungry, and I am succulent.

Awakened, heart shaking, I remind myself this is to be expected when you're expecting. Nearly 70 percent of pregnant women dream, and we dream more often and more memorably in pregnancy than at any other time in our lives. I'm amazed. As a kid, I kept a notepad at my bedside to try to capture my dreams, but they'd fade in the light. Now I have two or three dreams nightly, each as vivid and eerie as a midnight carnival. And all day long they haunt me.

The real spirits that inhabit me, of course, are hormones. Pregnant women are steeped in estrogen and progesterone, which fiddle with the dream phase of sleep known as rapid eye movement sleep (REM). Estrogen enhances REM sleep in humans. The longer we're in delicious REM, the more we dream. Progesterone lulls us to sleep, and the longer we sleep, the more we may dream. In the first two trimesters of pregnancy, we sleep up to two more hours nightly. We're cocooning.

Another reason we find our pregnancy dreams so vivid is that we simply have better dream recall, which comes from being awakened in the middle of a dream instead of having slept through it. Pregnant women often wake up several times a night, even in the first couple of months. I wake up from sweat and aches and a need to pee: sweat from fluctuating hormone levels, the aches of leg cramps and ligaments stretching, and a beat-up bladder to blame for the unwelcome bathroom break.

Another way to interrupt REM is to have an explosive 9.0 Richterscale orgasm. Seriously. This is another side effect of pregnancy: any dream may end in a nocturnal orgasm (or a fart). What happens is that during REM, the relative pulse pressure in the vagina increases, which is a trigger for orgasm. Meanwhile, estrogen lubricates the vagina, and the increased blood flow to the uterus and genitals can easily initiate arousal. Nearly 20 percent of women have more erotic dreams in pregnancy than they would usually, often about exes or possibly their obstetrician. Even asexual dreams are known to resolve in orgasm. In one of mine, I'm a ghost in a temple and I'm swinging a hammer. I wake up with a gasp, my uterus pulsing painfully. It's actually not sexy.

Of the pregnant women who dream, nearly 60 percent report that they have nightmares. Dreams mirror our inner conflicts and worries. And while the body is expanding, the mind is trying to keep up. According to psychologist Patricia Maybruck, author of *Pregnancy and Dreams*, there are seven categories of dreams: unresolved conflicts in childhood, fear for the baby's health, fear that our partner will no longer find us desirable, fear of childbirth, fear of lacking mothering skills, fear of loss of physical or emotional control, and financial stress. The unconscious is working the night shift. Interestingly, if you have a dream about the gender of your baby, you're more likely than by chance to be right, according to one small study at Johns Hopkins. Is there something the unconscious knows that the conscious mind doesn't?

Psychologists such as Patricia Garfield have found symbolic patterns in dreams. In the first trimester: water and swimming (representing amniotic fluid), carrying a heavy bag (feeling awkward), opening doors and falling or drowning and loss (fear of motherhood), buildings, construction sites, gardens and seeds (inner growth), and small aquatic animals like fish and lizards (awareness of embryo). In the second trimester: love affairs with former lovers (sexual desire and deprivation), husband having a love affair (insecurity), and dreams that involve one's own mother. In the third trimester: journeys (fear of unknown), large animals (awareness of moving fetus), and details of the labor and delivery.

But there's a bright side to all this nocturnal angst. According to Israeli psychologists Tamar Kron and Adi Brosh, dreamtime is when pregnant women do the heavy emotional work to prepare for labor and motherhood. In our dreams—or nightmares—we may resolve internal conflicts and address unconscious anxieties about the pregnancy. The purpose of dreams may be to help us process new information, which is why we dream more when our lives are changing.

Studies that support this theory are fascinating. Pregnant dreamers have a shorter labor than nondreamers—nearly an hour less, on average. Among the dreamers, those who had vivid nightmares had significantly faster deliveries than those who had pleasant dreams only. They also have a significantly decreased risk of postpartum depression. Perhaps women who don't have bad dreams are the type to repress their feelings. Later they find themselves emotionally unprepared for the shock of labor and motherhood. Are dreams the unreality that prepares us for a new reality that will seem unreal?

Looking back at my journal from my first trimester, I am struck by the patterns in my dreams. I take heart from last week's entry. I'm sitting on a rickety dock, playing river chess. The dock is a square called 5N, also inexplicably called 8T. I make a move and crash into the water.

But this isn't the end of the game. I bob! I feel the pull of the river as it moves to carry me downstream. I let it. I start to let go.

Do Fetuses Dream?

While you dream, so does your baby within. By the third trimester, fetuses sleep 85 to 90 percent of the time, cycling every twenty to forty minutes between REM and non-REM sleep. Starting around twenty-seven weeks, fetuses are likely to dream—and what they dream about is probably based on impressions of the outside world that they sense but cannot see. That poetry slam, the baseball game, the fight with your husband—the soundtrack of your life may echo on in your fetus's dreams.

For an unborn baby, dreamtime and wake time are probably very similar. When scientists tested preemies using an electroencephalogram, a tool to test the electrical signals produced by firing neurons, the brain in REM looked as if it were awake. From the fetus's perspective, dreamtime and wake time may be indistinguishable. In REM, the brain stem, the ancient reptilian region of the brain, takes over. It recycles waking hours by occasionally accessing parts of the neocortex where memories are stored and reprocessing them in ways we can't predict. For fetuses, REM may be necessary to complete the synaptic connections in the cortex, where conscious thought takes place. It promotes brain development. Dreaming is good for the baby's mind, rest assured.

2

SKINNY CHICKS, BOSSY BROADS, AND A BASKETBALL IN THE BELLY

The Biology of Boy-or-Girl

For weeks now, I have felt an occasional movement in my abdomen, pings and pops, which could be either the baby or gas. "Baby's kicks feel like fairy wings," a friend informed me in an excited whisper. The term for fetal movement is delightfully archaic: the quickening, as if a spirit is stirring. By now, week 19, about halfway though my pregnancy, the baby is becoming animated. It can taste, touch, hear, and kick. It's alive!

All along my baby's cells have been dividing, layers of tissue developing and folding inward and outward, origami-like, forming fissures and tubes, invaginating and gastrulating. Tissue expands and buckles, inflating like a soufflé. At this midway point, a fetus should be about 17 centimeters long and weigh 9 to 10 ounces. As the pregnancy books put it with their fruity analogies, the baby should now be a mango.

Frustrating for an information junkie like myself, the best tool to see our sweet fetus is ultrasound. Dating back to the 1950s, the ultrasound machine sends out sound waves and renders their echoes into fuzzy two-dimensional black-and-white images. It's adequate for the purposes of an anatomical exam: to measure the fetus's size, assess the position of the placenta and pelvic structure, and detect any major fetal abnormalities. (Some doctors use three-dimensional ultrasound, which uses the same scanning technology as two-dimensional but also has software to render the fetus in three dimensions and in color. A four-dimensional ultrasound is video.)

Most women opt for an anatomical ultrasound between weeks 18 and 22. If you haven't had amniocentesis, which tests the baby's chromosomes for abnormalities, this exam is a real highlight. It can reveal whether you're having a boy or a girl.

I have been at the edge of my seat waiting for this day, but it does not begin auspiciously. There's a blizzard outside, and I'm running late. I feel headachy. I put on a pair of low-heeled Mary Janes and take two steps out of the building. The sidewalk is slippery. I swivel around and run back upstairs to slip on a pair of sneakers. The clock is ticking. I sit on my office chair. The chair has wheels, and as I lean over to put on the sneakers, it rolls out from under me and I fall to the floor with a thud. Pregnancy, I learn the hard way, shifts your center of gravity.

All the way to the hospital I'm bedeviled by cramps and the queasy feeling that damage has been done. My husband meets me in the waiting room and tries to calm me.

"Where did you land?" he asks.

"On my butt."

"Ah!" he says cheerfully. "Well, then, the baby is fine." Lots of shock absorption there.

The ultrasound technician is more helpful. She reminds me that a nineteen-week-old fetus is suspended in a pool of amniotic fluid, like an egg in a waterbed. Pregnant women slip; it happens. A fall is riskiest in the third trimester, when there is not enough fluid suspension.

The cramps fade when I see our baby on the screen. She's not just alive; she's lively. This makes room for other anxieties to set in. The purpose of this appointment is to search for defects. Midway through a pregnancy, a fetus should have met various developmental milestones. Organogenesis, the development of the internal organs, is complete. The die is cast. What have we got?

As the technician probes my abdomen with an ultrasound transducer, a swirling chaos erupts on the overhead screen. She's jabbing here, prodding there, working her way around the fetus's tiny body. We watch fragments bob and sink in the cosmic undertow: a long ribbon of vertebrae, a beating heart, femur, tibia, humerus, ulna, radius, sternum, fingers. Then our first close-up: the head. It's 19 millimeters in circumference.

"Is that big?" I ask.

"Normal," the tech says.

The word detonates in the darkened room, followed by silence. We know fetal skull size has little to do with intelligence. But when discussing the baby's brain, we want to hear scintillating adjectives. Then shame hits me. Let's just hope the baby is healthy.

What I'd love to see, but can't with ultrasound, is the baby's brain under construction. New neurons are being generated at the rate of about 250,000 per minute during the nine months of gestation. This is astonishing growth; it's like one of those simulations of the big bang in fast motion. A whole universe busts out of the black.

The new neurons form extensive networks of connections called axons and dendrites. Axons are the superhighways of the brain; they carry signals long distance. Dendrites are the short-distance shortcuts; they receive signals and grow in earnest in the third trimester and after birth. Axons and dendrites resemble the limbs and twigs of trees, respectively. Dendrites reach out to one another, like God to Adam in Michelangelo's *The Creation of Adam*. There are neurotransmitters that carry signals across the chasm—dopamine, serotonin, acetylcholine—that regulate moods and metabolism. Some neurons are responsible for the baby's flutters, flips, and kicks. Others are behind vision, hearing, and touch. Yet others will shape appetite, both carnal and intellectual. At birth, the baby will have over 100 billion neurons. This number won't change much in her lifetime, but the number of connections between them will. Shortly after birth there will be a burst of growth, followed by pruning as she grows and learns. Chaos before order.

Then there's the heart. The tech apologizes for sinking the greasy transducer wand so deeply into my abdomen. She's angling for a better view of the four chambers. She applies more pressure. She tightens her lips. She zooms out and slides the wand to another location to try again.

It is now that we glimpse a vision that makes us coo: our tiny 10-inch, 10-ounce being raises its impossibly miniature hands and places them on top of its 19-millimeter head. We're smitten! I am astonished that an entity that did not exist twenty weeks ago seems to have the instinct to protect its little brain.

After five tense, heart-racing minutes viewing the fetus's heart—the four chambers and the valves that connect them—the tech smoothes her brow. The organ appears normal, she announces. And so, it seems, are the spine, stomach, kidneys, and placenta. She points out the three pulsing blood vessels in the sausage-like umbilical cord. The femur and thigh bone have been measured and declared normal. We all count ten fingers and ten toes.

Normal. I love this humdrum word.

And now the pièce de résistance: the baby's gender. The tech gives us a searching look. Do we *really* want to know?

Yes, we do. We really, really want to know. What have we got? A few clicks and zooms and it's ... three lines, two thick and one thin.

A doctor had told me to look for hamburgers or turtles. Three white bands, "hamburgers," representing the labia are girls, and three white dots, or a "turtle head" representing the penis and testicles, are boys.

We have a hamburger.

We gaze at Her in wonder.

> Prediction is very difficult, especially if it's
> about the future.
>
> —Niels Bohr

COULD GIRLS (OR BOYS) RUN IN SOME FAMILIES?

"It takes a man to make a man," a friend who sired three boys teased when we told him we're having a daughter.

"Actually," I tartly retort, "it takes a woman to make either." But the truth was, I was shaken. It took me days to accept that I'm having a girl. It's not that I don't want a daughter; I just had such a strong conviction about carrying a son. So much for a mother's intuition.

I'm not sure what made me feel so certain. Perhaps it's that my husband's four siblings are all male. His three first cousins are male. I figured Y sperm must be strong in his family. Males have proliferated in my family too. My first cousins, six of them, are all male; on both sides, I am my grandparents' only granddaughter. Boys run in both our families. Boys, boys, boys. Of course we'd have a boy.

Looked at mathematically, this isn't quite rational. If gender selection is roughly equal between the sexes, the chances of having a boy are one in two for a couple with two kids of the same gender; a one in four chance for three kids; one in eight for four kids; one in seventeen for five kids (such as my husband and his brothers); and so on. Statistically, there are families out there, even large ones, that have all boys or all girls by sheer chance. Sometimes you have to flip a coin over and over again before it comes up heads.

But logic isn't fun. The renegade in me seeks anomalies, quirks in the system. What happens if the coin is rigged to turn up tails more often, if only by dint of a decigram? That is, what if a man's X-bearing sperm, which would make a baby female, has a slight edge over his Y-bearing sperm, which would make the baby male, or vice versa? (Forgive me. I find this type of bias fascinating.)

From the very beginning, Xs and Ys don't always have the same opportunities. Is it shocking that males have a head start? We think the sex ratio is an even 50:50 split between boys and girls. Turns out, this is close to the truth, but it's not the whole truth. About 106 boys are born for every 100 girls in most parts of the world. Although most men have an equal number of Xs and Ys, Y sperm may make it to the egg sooner because they are faster and lighter than X sperm, giving males an edge. More male embryos survive the earliest stages before implantation. (The ratio of boys to girls may be as high as 130:100 at conception.) According to evolutionary theory, there are more males than females born because the surplus dwindles after birth. Males are actually the weaker sex. Boys are hit harder than girls—from birth defects to accidents. The sex ratio evens out in adulthood. In old age, widowers often get their pick of dates because they're far outnumbered by widows.

Yet there are straightforward reasons that a man might be more likely to have daughters than sons. A notable one is long-term exposure to reproductive toxins: pesticides, fungicides, fumigants, dioxins, lead, and other nasties. Cancer survivors who have had chemotherapy have more daughters, and so do long-term smokers and heavy drinkers. Some toxins alter hormone levels, decreasing testosterone and increasing progesterone, which may indirectly favor X-bearing sperm. The more likely culprit is a permanent genetic mutation that causes the defect.

While exposure to a toxin may damage a man's X- and Y-bearing sperm equally, it's the male embryos that take the hit harder. This is because daughters have two X chromosomes (XX), one from each parent, while sons have only one (XY). Daughters have a backup X chromosome, and sons do not. If something goes wrong with a male's only X, he's in trouble. Many male embryos are miscarried, usually so early in the pregnancy that a woman might think she's just having a late period.

Genetic mutations may explain why some families have all girls. But where do surpluses of boys come from? What about all the Mitt Romneys and David Beckhams who sire boy after boy after boy (five and three, respectively)? Coincidence—or not?

A sex-selecting gene explains a boy bias, according to Corry Gellatly, a research scientist at Newcastle University in Great Britain. Analyzing 927 family trees containing birth information on over a half-million people dating back to the time of Shakespeare, Gellatly came to a striking statistical insight: men with more brothers are more likely to have sons, and men with more sisters are more likely to have daughters. This pattern is found in men but not in women. It doesn't matter if boys run on your mother's side of the family. Only the father's side counts.

While both parents are likely to have the sex-selecting gene that has yet to be identified, only males pass it on because only males produce sperm. About 20 percent of men have a sex-selecting bias, Gellatly estimates. I am astonished. One out of every five men! And among that 20 percent, men with boy-biased genes produce more Y-bearing sperm, and men with girl-biased genes produce more X-sperm. Gellatly believes there are currently more men with a male bias, which explains why the sex ratio at birth is often at least 106 boys per 100 girls.

To explain this intriguing phenomenon, Gellatly uses sweeping evolutionary terms. This isn't about picking colors for the nursery; it's about the survival of the species. At times when there are too many males in a population, men with boy-biased sperm (families with mostly boys) have fewer chances to pass on their genes to future generations than men with girl-biased sperm (families with mostly daughters) because the males (in general) will have fewer mating opportunities than females. The opposite happens when there are too many females. Populations seesaw back and forth maintaining the gender balance. Another way of thinking about sex-biased sperm is that it's a biological reaction to population changes. If a war wiped out much of the male population, then men with boy-biased genes would help restore the male population within a few generations. This is because families with many sons are more likely to have at least one boy who survives the war, and this son may very well carry his father's boy-bias gene. This would solve the mystery of why, after wars (including both world wars), more sons have been born than daughters.

I'm deeply curious about whether we're close to finding the sex-selecting gene. "I think it is quite possible that the gene will be located,"

Gellatly says. "But the way things are at the moment, it will probably require an economic or medical incentive for the research to be funded." Research on fruit flies has found genetic effects on the sex ratio, but human studies are not yet in the works. Science is rarely funded for curiosity's sake alone. Truth is, many people are comfortable with the idea that chance is solely responsible for the gender of their baby. And most of the time, they're right.

———

Next is the mother's contribution to the sex of the baby. Many people mistakenly think it's Dad who solely decides whether Baby will be male or female. This is technically true—it all depends on whether a paternal X or Y chromosome fertilizes the egg—but Mom can influence which sperm gets to the egg and whether an embryo survives. Instead of tweaking the sperm directly, we influence the environment in which the sperm and the fertilized egg try to survive. After all, we *are* this environment. We are the stream to swim through, the shell to penetrate, the sea to float to, and the sand to burrow in. Sometimes the environment favors Xs, and sometimes it favors Ys. The contest for survival of the fittest takes place in the arena that is us.

When a woman has a biological sex bias, it often has to do with her condition at a certain moment. We're not fixed. Our conditions change. Our cycles shift. Our hormones rise and fall. Our weight can go up and down. Our body chemistry fluctuates yearly, monthly, daily, and even hourly. All this may affect the gender of the baby. The tiny being we are creating is a product of both our inner and outer worlds: the times we live in, the stresses we're under, the food we eat, our status in society. These forces are evolutionary. They're ancient; they precede us. And as we shall see, they were once meant to preserve us.

> If we were living in ancient Rome or Greece,
> I would be considered sickly and unattractive.
> —Gwyneth Paltrow

DO SKINNY CHICKS HAVE MORE DAUGHTERS?

In the fall of 1970, two men met each other in a primate behavior class at Harvard University. One was Dan Willard, a slim mathematics graduate student who wanted to meet women. A course about animal behavior, Willard thought, would attract coeds interested in men. The second man, Robert Trivers, was a six-foot-two graduate teaching assistant with a tendency to swear and spout out provocative theories. In one of his lectures, Trivers made a point that in the human species, which is more or less monogamous, women tend to marry up on the socioeconomic scale. A poor girl is more likely than a poor boy to grow up and marry well.

Willard contemplated this and had a brainstorm. Wouldn't it make sense, he asked Trivers, for rich parents to have more boys to take advantage of all the social-climbing girls and for poor parents to have more girls who could marry up? Even more provocative: Are humans already doing this without even knowing it?

Trivers, who would become a famous evolutionary biologist, loved the suggestion. While Willard returned to his mathematical and romantic pursuits, Trivers decided to research the idea some more and broaden it. Over the next three years, he conceived a theory that forever changed biology: parental investment theory. The theory, in a nutshell, is that babies are costly, especially for the mother. To get the most from their investment, mothers should favor the sex that is more likely to pass along her genes to the next generation. When food is plenty and conditions are good for the mother, sons are the better investment because a strong, healthy, well-fed male will attract many mates. When conditions are not so good, females are the better investment, because an underfed daughter is more likely to attract mates than a malnourished, shorter-than-average, and/or low-status son. Wherever Trivers looked, he found evidence that this bias really does happen in the real world—in deer, rabbits, mice, and even birds and bees. Among fellow evolutionary biologists, this trend has become known as the Trivers-Willard effect.

Soon enough, researchers found evidence of the Trivers-Willard effect in humans, too. One marker of resource scarcity they've looked for is a woman's weight before pregnancy. Researchers in Italy collected data on nearly ten thousand new mothers and found that those in the lightest twenty-fifth percentile—women who weighed 119 pounds or less before pregnancy regardless of height—gave birth to significantly more daughters than did women who weighed more (51 percent versus 47 to 48 percent in the higher quartiles). If a woman is extremely thin, her body may interpret caloric restriction as a sign of stress and scarcity. This may or may not be true today, but our bodies don't know that. Skinniness once meant famine, not fashion.

Over time, the evidence that skinny women have more daughters grew. In Ethiopia, another team of researchers collected data on women who had recently given birth. In this study, skinniness was determined by the mid-upper-arm muscle-area index; considered a more accurate measurement of nutritional status than body mass, it's the circumference of the upper arm. Women with limbs like Kate Moss's, in the lowest twenty-fifth percentile, were more than twice as likely to have recently given birth to a daughter than their stronger and heavier peers in the top twenty-fifth percentile. Meanwhile, a team in Norway followed nearly forty thousand women during and after their pregnancies. They found that women with anorexia before pregnancy were 10 percent more likely to have daughters than average-weight women. Those with bulimia were 9 percent more likely to have daughters.

The Trivers-Willard effect may also apply in a more universal way to explain intriguing findings that link hardship or health and gender. Married moms or moms living with their partners tend to have more sons, and single moms have more daughters. Older mothers have more daughters, and younger women have more sons. Women in rich nations have more sons on average; women in poor nations have more daughters. East Germans gave birth to fewer boys before the fall of the Berlin Wall, and so did New Yorkers nine months after the fall of the World Trade Center towers. Big and tall parents have more sons; petite parents have more daughters. Women who marry billionaires have more boys: a spanking 60 percent of births are male, according to a study based on the 2008 *Forbes* billionaire list. (Interestingly, this doesn't hold true for self-made female billionaires.)

So here's the billion-dollar question: How do our bodies discriminate between sons and daughters? There are no definite answers, but the most popular theory is that undernourishment makes the womb less accommodating to the male embryo, which is weaker and needier than the female. (Make of that what you will.) If you eat sparingly, have a low-fat, low-calorie diet, or are very stressed, your glucose (blood sugar) levels are likely to drop below a critical threshold. Low glucose levels are associated with a deficiency of progesterone, the hormone that thickens and maintains the lining of the uterus. This is bad news for male embryos. When the womb isn't richly lined, the hardier female is more likely than the male to attach, implant, and sustain itself. Girls can make do with less.

How do we know that this happens? Diabetic women who take insulin, which causes low glucose levels, have significantly more daughters than sons. But hop up a woman on sugars, as has been shown by in vitro fertilization treatments that culture fertilized eggs in a high-glucose medium, and more male than female embryos thrive. In their study "You Are What Your Mother Eats," researchers at the universities of Oxford and Exeter revealed that among women who ate a high-calorie diet before conception (2413 kilocalories daily on average), 56 percent gave birth to boys, compared to only 45 percent among women who ate a somewhat lower-calorie diet (2283 kilocalories daily). Academics debate whether it's a random association, but the one food particularly related to male births was breakfast cereal. Women who ate a bowl every morning were nearly twice as likely to have sons as did breakfast skippers. Skipping breakfast, the researchers suggest, "lowers glucose levels for long periods of time, and may be interpreted by the body as indicative of poor environmental conditions."

The sex bias is slight—starving yourself until the buzzards follow you won't guarantee a daughter—but it's consistent and statistically significant. It's unknown whether there's a difference in female births between women who are naturally beanpole thin and those running a nutritional deficit due to dieting or circumstance. Chronic undernutrition may have a different effect from sudden starvation. When the Dutch suffered an abrupt winter-long wartime famine in the mid-1940s, no more girls than boys were born later. Of course, dieting to conceive daughters is silly, and if you waste away during pregnancy, you might harm your baby's, and even your future grandchildren's, health (see page 62). The Trivers-Willard effect is just one of those evolutionary quirks, and like anything else applied to individuals rather than a huge population, it's highly unreliable.

An evolutionary bias for females is now irrelevant in much of the world where thinness represents abundance more than scarcity. If scrawny starlets and socialites had more daughters, those princesses would grow up pampered, not paupered. The only problem is that they might complain about a lack of suitable bachelors in their social set. How many billionaire boys will there be?

Double X or Bust?

In the 1960s, the biologist Landrum Shettles claimed to have a scientific solution to the age-old quest to select a baby's sex: know when you're ovulating. The concept behind this is that cervical mucus becomes very slippery when we ovulate (and orgasm). Y-bearing sperm take less time to slalom through the vagina than X-bearing sperm, which are bigger and heavier because they contain more DNA. Xs are turtles, Ys are hares. In the race to the egg, the sprightly Ys would get to the finish line faster, but the Xs would last longer. This means that if intercourse happens on the day of ovulation, we're more likely to have a boy. But if intercourse happens two to four days before ovulation, more Xs survive the long wait and the harsh preovulatory conditions, and a baby girl is more probable. The timing of intercourse may also explain why more boys are born to women who ovulate early in the cycle or have a follicular phase of fifteen or fewer days between period and ovulation, and more girls are born to women whose follicular phase is seventeen days or longer.

But does the Shettles method work? Several scientific studies (by researchers other than Shettles) say no: the timing of sexual intercourse

In relation to ovulation bears no statistical significance on the sex of the baby. I suspect the Shettles method offers a real bias, but it fails because it's just not practical or possible for most couples to predict the exact timing of ovulation, not even with a kit. Suffice it to say that for every Shettles success story, there's another that says the method doesn't work. But at least there's fun in trying.

The best chance to clinch a baby's gender is through a sex preselection process. Most techniques rely on the fact that Xs are heavier than Ys and therefore travel through layers of a gel more slowly over a period of time. Xs and Ys can be sorted and separated, with a success rate of about 70 percent. A more high-tech process is known as fluorescence in situ hybridization. DNA probes are introduced to the semen, one for Xs and one for Ys, and each gives off a different neon color when it binds to the chromosome. An embryo of the desired gender is then transferred to the woman's uterus as part of the process of in vitro fertilization. Sexy it isn't, but it's the most accurate way to choose the baby's sex.

DO BOSSY BROADS HAVE MORE SONS?

To the couple hoping for a boy, the traditional advice is to eat red meat, let the man initiate sex, and make sure he climaxes first. Sexist and old-fashioned, it's the "it takes a man to make a man" mind-set.

So persistent is the belief in the masculine begetting the masculine that it has made its way into modern science in the form of another popular theory: that the hormone testosterone has an effect on the sex ratio, favoring boys. Testosterone is normally associated with males. And not just any males but alpha males—high-powered, dominant, macho, aggressive types. There's little proof that brutes really have more boys, although some studies suggest that men with abnormally low testosterone levels may have more daughters.

But it's a misconception that testosterone is only a he-hormone. Testosterone runs through women's veins too, albeit at serum levels that are about ten times lower than men's. That's an average. Some women are born with naturally high testosterone levels—so high that they overlap the lower end of the male range—or have circumstances that raise their levels. In both sexes, high testosterone levels are related to behavioral and personality differences in dominance. Call a high-testosterone woman what you will—bitch, butch, bossy, bully. She is confident, competitive, forceful, edgy, ambitious, and more likely to have a high social status.

In this context, it's not sexist when Valerie Grant, an evolutionary psychologist at the University of Auckland and mother of four (boys), steps up and asserts that high maternal testosterone is linked with male babies. Grant's research began over forty years ago, when she said she could intuitively guess the gender of unborn babies based on the personalities of their pregnant mothers. Her competitive, domineering friends tended to have boys, and her shyer, compassionate friends tended to have girls. In the 1980s, she created a test to measure person-ality dominance. Administered to women before and after they became pregnant, the test asked how often they experienced various feelings: influential, proud, sheltered, humiliated, admired, useful, shy, guilty, and so on. The more confident and assertive the woman, the higher she scored on the dominance test. The higher a woman scored in dom-inance, the more likely she was to have a boy, or all boys if she had more than one child. (An online version of Grant's test is available at www .sexratio.com/test. My results predicted an 85 percent chance of a girl. Hmmm.)

As the years passed, other researchers offered similar observations. Satoshi Kanazawa, an evolutionary psychologist at the London School of Economics, published a study suggesting that hormones play a role here: women with strong "male brains," such as scientists, engineers, and mathematicians, have more sons than women with strong "female brains," such as nurses, therapists, and schoolteachers. J. T. Manning, a biologist at the University of Liverpool, found that women (and men) who were exposed to high levels of testosterone in the womb are more likely to have sons. (Evidence of high fetal testosterone levels can be measured by the ratio of index to ring fingers; see page 155.)

Focusing on testosterone, Grant developed what she calls the maternal dominance hypothesis: a woman conceives an infant of the sex she is psychologically suited to raise in her current condition and social environment. "Whatever sex infant a woman conceives is the right and best one for her at that time and in that place," wrote Grant. Dominant mothers may be more initiating, which suits boys, and empathetic mothers may be more responsive, which suits girls.

The maternal dominance theory is a generalization of the Trivers-Willard effect: offspring are born into their mother's environment and social situation, so the sex that thrives better under such conditions is the favored one. In this case, it can be argued that dominant women are more likely to be competitive and have a higher social status, which would benefit sons more than daughters; at least it did in the ancestral environment. A competitive, high-status male would have more mating opportunities than a passive, low-status male, while a female is more likely to find a mate regardless.

Grant's hunch is that preferential treatment to one sex or the other happens at the moment of conception, when sperm are swarming around the egg. It's difficult to ask women to donate their eggs to research, so she and her colleagues recruited another female mammal: the cow. As with humans, cow eggs develop in the ovaries. Each egg is suspended in a bubble called a follicle, and each follicle is filled with fluid. Follicular fluid happens to contain testosterone. As Grant suspected, eggs that had been marinating in high concentrations of testosterone for eight weeks prior to conception were more likely to be receptive to Y-bearing sperm when fertilized in vitro, and more males were conceived. Eggs that had developed in low concentrations of testosterone were more receptive to X-bearing sperm, and more females were conceived. It's as if eggs play favorites.

Testosterone levels fluctuate all the time. While some women are consistently high or low in testosterone over months or years, external factors—a competitive situation at work, a family feud, a move to a new home, even a sudden stressful encounter—can temporarily raise our levels. Testosterone levels fluctuate over the years, usually decreasing as we age. Grant estimates that most of us—about 68 percent—have testosterone levels that fluctuate within one standard deviation of the average and over a period of years may have children of either sex, while about 16 percent of us with much-higher-than-average testosterone consistently have a boy bias and 16 percent with much-lower-than-average testosterone consistently have a girl bias.

The maternal dominance theory remains controversial. It's hard to quantify: Exactly how high do testosterone levels need to be to affect our eggs? How does testosterone in our blood correlate with testosterone in our egg follicles, and what personality traits or circumstances correspond with those levels? Can we isolate what it takes to raise our testosterone level enough for a boy bias or lower it enough for a girl bias? It's complicated.

Incidentally, a study of the 2008 *Forbes* billionaire list found that self-made billionaire women tend to have more girls than boys. This seems to contradict the maternal dominance theory (although women who married money or inherited it had more boys). Why would women billionaires have more daughters? The sample is too small to be significant, but it makes one wonder if evolutionary influences are weakening. In the twenty-first century and beyond, why wouldn't daughters benefit as much as sons from their mom's high-testosterone status? Or why wouldn't sons benefit from their mom's low-testosterone status?

It may turn out that empathic, compassionate, low-testosterone men will be increasingly valued in the modern world. After all, those bossy bitches out there will need someone to help raise the kids.

The Truth about Basketballs and Watermelons

My bump rides high and tight. It's all in the front. Looking at my back, no one would think I'm pregnant. This means I'm carrying a basketball. Strangers on the street see this as an omen. The basketball is a boy, I'm told. Boys are basketballs. Girls are watermelons. A watermelon-bearer carries low and her load spills over the sides.

Well, they were wrong about me. But is there even an inkling of truth to the theory? Janet DiPietro, a developmental psychologist at Johns Hopkins School of Public Health, asked the same question. Recruiting over one hundred women who did not know the sex of their baby, she asked them to make an educated guess. Among those who guessed based on the appearance of their bump—whether they were carrying basketballs or watermelons—accuracy was lower than by chance, about 43 percent.

Interestingly, a guess based on feelings or dreams was much more accurate, about 71 percent. Expectant moms with a minimum of a college education made more accurate guesses about the gender of their babies than did women with high school diplomas only. Incidentally, none of the other popular methods of gender prediction have passed the scientific smell test. Predictions based on the Chinese birth calendar (gender determined by month and year of birth) and fetal heartbeat (based on heart rate, higher than 140 beats per minute for girl and lower for a boy) bombed in studies.

The way we're carrying has nothing to do with the gender of our baby, but it does reveal something about our abdominal muscles. First-time mothers and fitness buffs tend to have "basketballs" because their taut muscles support the baby higher. "Watermelon" bearers, usually veteran mothers, have weaker abdominals and therefore carry lower. The upside is they breathe and eat more easily, leaving more space for lungs and stomach (albeit less for the bladder). Basketballs can turn into watermelons over the course of the pregnancy as the baby grows and ripens.

Surprisingly, your shape on top is a more accurate gender predictor. Women carrying girls grow larger breasts during pregnancy than women carrying boys (8 versus 6.3 centimeters on average). Male fetuses produce more testosterone and require more energy—conditions that may suppress breast tissue growth.

> Pink isn't just a color, it's an attitude too!
>
> —Miley Cyrus

WHY WOULD GIRLS PREFER PINK?

Now that we're having a girl, I'm bombarded by pink: pink rattles, pink nappies, pink onesies, pink doggies. I'm gifted a Pepto Bismol–colored maternity shirt that reads "Tickled Pink." Honestly, all this pink makes me a little sick to my stomach.

Pink for Girls, Blue for Boys wasn't always so. Once upon a time, blue was considered the weaker, and therefore feminine, color. Pink, a starter shade of mighty red, was appropriate for a little man. Sometime in the nineteenth or early twentieth century, the custom reversed, and pink became the customary color for baby girls, blue for baby boys. Seen this way, gender-specific colors would appear to be entirely cultural.

The story could end there, but several studies have found that women genuinely seem to prefer pinks and reds more than boys and men do (they're twice as likely, according to one finding), and some researchers think biology has as much to do with it as culture. An in-depth experiment at the University of Newcastle found that women in China who were raised without a pink passion—no Barbies and ballerinas—preferred shades of red (especially reddish-purplish ones) just as much as or more so than did British women. Among the two hundred or so participants in their twenties, both groups of women, the Chinese and the English, preferred colors with a reddish contrast more than did their male counterparts, who preferred blue-greens. To the researchers, these results suggest an underlying biological basis to women's bias for shades of red, including pinks and purples.

One explanation is that women are more sensitive to shades of red than men are because women once specialized in foraging. Our foremothers needed to see ripe reddish fruit among the green bushes and brambles. Males, the hunters, didn't need to be as perceptive. (Nor do carnivorous animals; they're red-green color-blind.) The ability to distinguish subtle shades of red is particularly useful to women in other ways too, because so much social information—being furious or feverish, lying or being shy—is expressed in these hues. Women are often better than men at picking up on emotional nuances.

By itself, the theory that women prefer reddish colors because they're more sensitive to them isn't entirely convincing, but it's backed by a biological argument. It goes like this: We all have color detectors in our eyes called cones. Cones contain red, green, and blue pigment-sensitive molecules called opsins. It turns out the genes that make red and green opsins are on the X chromosome (blue opsin is on its own, sitting on a nonsex chromosome). As we know, men have only one X chromosome, whereas women have two—one from each parent. If men inherit a defect in the red or green opsin gene they inherited from their mom, they don't have another X chromosome as a backup, so they get red-green color blindness. Women are covered because they get opsin genes on both X chromosomes. In female eyes, some cells have the mother's opsin gene activated, and others have the father's opsin activated.

Now comes the interesting part. Women with their two X chromosomes may inherit extra opsin genes, with not just three but four different cone receptors: blue, green, red, plus a slightly shifted red (or green). Given how easily mutations occur on this part of the X chromosome, extra cones are no freak occurrence: an estimated 15 to 50 percent of women have a slightly shifted fourth type of opsin gene. This means that one type of red cone is activated on one X chromosome, and the other type of red cone is activated on the other X chromosome. While a normal three-cone carrier (a man) can detect about 1 million different shades of color, a woman with four cones could theoretically make out up to 100 million different shades, depending on how little overlap there is between their two red (or green) cones. This superwoman would, for instance, have a richer red sensitivity—detecting the finest nuances of salmon, cerise, carmine, coral, puce, lavender, magenta, scarlet, rose, and so on. If you're female, there's a fair chance you have a supersensory ability to see reddish hues, including pinks, but don't know it.

People with an enhanced sense of taste tend to prefer foods with subtle flavors. Does an enhanced ability to see red result in a preference for the nuanced shades of red? That remains to be seen—but you can't prefer a color if you aren't sensitive to it.

Despite all her bubblegum-colored baubles, my daughter, at least in her infancy, is unlikely to prefer pinks over other colors. Although all four- to five-month-olds gaze longest at reddish hues, infant girls don't appear to prefer them more than infant boys do. (There's much speculation on how long it takes for children to detect a full color range. Some researchers say at least a year.) And even if a girl grows up pink preferring, the color may not be her favorite as an adult. As the University of Newcastle researchers point out in their cross-cultural study, "Girls' preference for pink may have evolved *on top of* a natural universal preference for *blue*" (emphasis mine).

Yes, blue. Many color preference studies show that men and women prefer blues most of all, albeit with women preferring reddish-blues like purple and men preferring greenish-blues like teal.

So let's strike a compromise. Pretty in pink, prettier in purple.

3

FAT RATS, SALMON-AND-SOY SUPPERS, AND WHY GRANDMA'S DIET MATTERS

Food and the Fetus

At twenty weeks and two days pregnant, we (the fetus and I) are eating pound cake. Pound cake is what I have been craving, along with other simple carbohydrates. Pound cake, I tell myself, has eggs in it, six to be precise, and eggs are a source of calcium and choline and other nutrients. These eggs will find their way into my baby girl's body and brain.

So what is my little girl made of? Sugar and spice and butter and eggs. Thinking about how my diet affects my unborn baby, I'm reminded of a line from a Walt Whitman poem: "There was a child went forth every day, / And the first object he look'd upon, that object he became." Spinach, chips, chocolate, licorice. She is the oranges and dark chocolate we ravaged over the holidays and my favorite apricot tea. She is the vitamin B complex I faithfully took in the first trimester. She is the guilty pizza binges, the crazy carrot habit, the sanctimonious salmon suppers. She is what I eat, and she is what my food eats. She is rainfall and grass. She is sunshine.

Whatever I do affects this self that is in me and of me but not me. As an entity, I now have two brains, two mouths, two tongues, two hearts, four eyes, and eight limbs. "I love you, both of you," my husband murmured in bed last night. I stayed awake for a while after that, lying in the dark and listening to the whir of the ceiling fan.

So entwined are we, the fetus and I, that everything I eat, drink, do, and feel may affect the behavior of her genes. The field of science that explains how this happens is called epigenetics. Diet, hormones, toxins, stress, viruses, and even certain experiences in a mother's life may epigenetically alter the recipes of her baby's genes—meaning the baby inherits genes and the modifications that overlay them. Epigenetics gives us a fascinating new way of looking at our lineages and legacies.

The story of the honeybee describes the power of epigenetics. In a bee colony, all females have the same DNA. They're clones—until one grows bigger than all the others and becomes the queen. She has plumper ovaries and a life span measured in years, not weeks—only because she gorged on something extraordinary. Back when she was a young larva, worker bees souped-up her cell with lavish amounts of royal jelly, a substance high in B-complex vitamins. It turns out that binging on royal jelly inhibits the activity of a gene that turns most honeybees into workers and not royal highnesses. At least in the bee kingdom, it's not genes that make queens. It's diet that changes the way genes behave, and any bee has the potential to become royalty.

———————

It's not easy being at the top of the food chain. At my last checkup I heaved myself onto the scale at the doctor's office. The nurse did a double take as she moved the slider across the beam, notch by notch, Realizing she needed the larger weight to balance the bar, she shot me an amused look. "No more vomiting, eh? Looks like you're keeping it all down."

The average woman carrying a single fetus gains twenty-five to thirty-five pounds during pregnancy: three to five pounds in the first trimester and one to two pounds per week in the second and third trimesters. Where is all the weight going? Two pounds to the placenta; six to fluids, both blood and amniotic; two to the ballooning uterus; two to the swelling bosom; and another seven pounds to fat stores in the thighs and butt. Seven to eight pounds should eventually go to the baby.

All that pound cake goes into one seven-pound baby.

Let there be such oneness between us, that
when one cries, the other tastes salt.

—Anonymous

CAN MOM'S DIET AFFECT BABY'S GENES?

If my baby's genome were a cake recipe, then I'd be tweaking it subtly. I'm saying something like slip in more egg yolk. Double the lemon zest. Bake ten minutes less. My changes are like directions on sticky notes. The master recipe is intact. Nothing has been erased or permanently altered; the ingredients are the same. But the final product is different from what it would have been under other circumstances.

Tweaking the genomic "recipe" is what epigenetics—altering the activity of genes without changing their underlying DNA sequence—is all about. In technical terms, the foods we eat contain organic compounds called methyl groups that attach themselves to our unborn baby's genes and change their behavior, especially early in development. Adding methyl groups turns gene expression down (methylates) and removing them turns it up (demethylates), like turning the volume higher or lower, on or off. The result can be as dramatic as slightly altering a recipe for a puffy, bright-yellow cake so that it comes out dense and nut brown.

In a way, that's exactly what happened when scientists at Duke University and Baylor College of Medicine experimented on a species of mouse called agouti. The photos accompanying their study show two agouti mice, one obese and bright yellow (considered normal for this lab breed) and the other compact and nut brown. As radically different in appearance as the two rodents were, they were as genetically similar as identical twins. If you looked at their DNA alone, you'd see the exact same code in both. The only difference between them is that two weeks before conception and throughout pregnancy and lactation, the mother of the compact brown mouse ate chow supplemented with chemicals that donate methyl groups to genes—in this case, B-complex vitamins such as vitamin B_{12}, folic acid, choline, and betaine. Early in the embryonic development of the brown mouse, those methyl groups attached themselves near the sites of genes that normally promote yellow pigment and obesity (linked to diabetes and cancer) in agouti mice and silenced them.

It was an astonishing experiment. Here was an animal whose predisposition to obesity, diabetes, cancer, and yellow fur was overcome by maternal diet alone. The mother was fed only three times as many supplements as found in normal mouse chow—the human equivalent of an extra dose of vitamins. Foods that contain these nutrients are meat, liver, shellfish, milk (for vitamin B_{12}), leafy veggies, sunflower seeds, baker's yeast (for folic acid), egg yolk, liver, soy (for vitamin B_6 and choline), and beets, wheat, spinach, and shellfish (for betaine).

Later studies found that genistein, a weak estrogenic chemical found in soy products, had the same Cinderella effect, transforming homely mice with a genetic predisposition to obesity and cancer into stunningly healthy brown ones. Isn't it interesting, the researchers point out, that Asians, the population that consumes the most soy in the world—2,050 times more per capita than in the West—have the lowest risk of breast, colon, and prostate cancers? And isn't it uncanny that Asians conceived and raised in the East enjoy lifelong protection from breast cancer while their daughters, raised on a Western diet, do not? Genistein in soy seems to prevent genes from overexpressing themselves and causing cancers. This reminds me of what we say about kids when they act up. It's not the genes themselves that are bad, *it's the way they behave*. And diet affects a gene's behavior.

Parents who are yearning for a baby genius may be intrigued by choline, another B complex vitamin that is essential for brain development. A methyl-bearing chemical found in eggs, liver, salmon, and organ meats, choline has epigenetic properties too. When expectant mother rats were fed four times their usual intake of choline in late pregnancy, their offspring had fewer tumors than rats that didn't eat it. The pups of choline-binging mothers had visual and auditory memories vastly superior to those of pups whose mothers ate a normal diet.

In old age, these brainy rats also escaped the usual fate of rodent senility. Choline appears to affect the behavior of genes that alter the timing, placement, and behavior of neurons in the hippocampus, the memory center of the brain. In humans, development of the hippocampus may continue for years after birth, suggesting that we should stuff ourselves with eggs not only when we're pregnant but also if we're breast feeding. Two eggs daily almost satisfies the dietary requirement of 450 milligrams in pregnancy (550 milligrams when nursing). Did Einstein's mom love omelettes?

The natural temptation is to gorge on choline, folic acid, soy, and other methyl-bearing supplements. We shouldn't. Getting these nutrients in the foods we eat is beneficial; excessive pill popping is not (although folic acid supplements are often prescribed by doctors because of the widespread deficiencies in B vitamins). Throughout fetal development, genes methylate and demethylate all by themselves, turning off and on, higher and lower. This is why heart cells behave differently from brain cells even though both have the same DNA. Timing is everything to the fetus. At an early phase of development, it might be helpful for the baby to suppress or overexpress a certain gene, but the same tweaks could be harmful later. Add extra sugar at the beginning of the recipe and the cake tastes better; add it later and the texture goes to hell. Some of us may have underlying genes that are supersensitive to small quantities of certain supplements—just as in some recipes, even a pinch of chili pepper goes too far. Quantity counts too. We can expose the fetus or ourselves to too much of a good thing.

For instance, studies on folic acid found that a minimum of 4 milligrams daily dramatically reduces the risk of neural tube defects—disorders in which the neural tube fails to close, resulting in spina bifida (an opening in the spine that may cause paralysis) or anencephaly (an opening near the head resulting in an absence of much of the brain and skull), among other conditions. Preventing these birth defects is why fortified bread and cereal are a great public health coup, and it's why doctors often prescribe folic acid pills—because some of us don't eat enough leafy green vegetables, beans, egg yolk, and liver. But the big picture is more complicated. There is evidence that excessive, overexuberant supplementation of folic acid and other B vitamins during pregnancy is linked to higher rates of asthma (in human babies) and tumor development (in mice).

Some experts fear that excess prenatal multivitamin consumption, acting through epigenetic factors, may predispose children to the curses of obesity, heart disease, and type 2 diabetes later in life. When pregnant rats on an otherwise normal diet were overfed multivitamins, their babies grew up hungrier and with little ability to control their food intake. Expectant moms exposed to a glut of vitamin E in supplements gave birth to babies with a five- to ninefold increase in risk of congenital heart disease. Consider the epidemics of obesity, heart disease, and other medical problems in the United States, and the news is especially thought provoking. The health-obsessed (and the nutritionally lax) take two to seven times the recommended dietary allowance of ten vitamins, not to mention fortified cereals, energy bars, and vitamin-enriched water.

Not even soy is above suspicion. Researchers suspect there are phases in fetal development when exposure to a lot of genistein in soy is protective against cancer and times when it is not. One study found that fetal exposure to soy must be followed up after birth and into adolescence for true cancer protective benefits. But it's not clear-cut. Some researchers warn that too much soy too early in life, as when newborns drink soy-based infant formula, may increase cancer risk because it interferes with other developmental processes, or because it's consumed with folic acid, which may cause conflicting epigenetic effects.

These studies on soy, choline, B_{12}, and folate give us just a taste of the chemicals in foods that alter gene expression. Targets of future study include epigallocatechin (in green tea), baicalein (in a Chinese herb) and curcumin (in curry), ibuprofen, lycopene (in tomatoes), pomegranate extracts, quercetin (in plants), selenium (in plants and seafood), alpha- and gamma-tocopherols (antioxidants), valproic acid, and zinc.

Vitamin D (in eggs, yogurt, wild salmon, liver, and sunlight) is thought to be an epigenetic gold mine. A prenatal vitamin D deficiency is associated with a higher risk of asthma and wheezing in childhood.

Omega-3 (in fish and flaxseed) affects genes involved in brain development (see page 64). Alcohol and a high-fat diet may increase the risk of disease by turning down or switching off vital genes, while a healthy diet may decrease or reverse the risk. Some foods may interfere with another's health benefits; for example, green tea reduces the absorption of folic acid.

How simple it was for scientists to once think our and our babies' genes were closed systems, separate and removed from drinking water, hen's eggs, chocolate-covered candies, and all else they may encounter! Now we know, in a very real sense, that genetic behavior depends on our eating habits.

Epigenetics is cutting-edge, but the prenatal diet is not. It's essentially the one we were taught as schoolchildren: eat leafy greens, eggs, fruit, fish, nuts, beans, whole grains, soy, and some fat and dairy. Because excess pesticide exposure can also harm a fetus's development, we should scrub our fruits and vegetables and try to eat organic varieties of those that are the most contaminated: peaches, apples, sweet bell peppers, celery, strawberries, cherries, pears, grapes, spinach, lettuce, and potatoes. Everything low-toxin, nutritionally balanced, and in moderation. Some cake is okay, not too much (but how could you eat too much cake?).

This is all the practical wisdom that science can offer—at least until diet can be tweaked for each mother's and baby's genome. There isn't a perfect recipe, not yet anyway, for the bun in the oven.

The Badness of BPA

Could a good diet offset the poisons we unknowingly consume? The latest research indicates that nearly all of us are loaded with bisphenol A (BPA), a chemical used to harden plastic. BPA leaches out of plastic toys, tubing, thermal paper used for receipts, and food packaging. Canned tomatoes are especially steeped in the chemical because acids degrade the lining of the cans. BPA is found in significant quantities in our bloodstream. It's also in the placenta, amniotic fluid, and breast milk. It's in the fetus too, potentially leading to prematurity, miscarriage, birth defects, and other more subtle developmental problems. Only recently have some states banned the use of BPA in the manufacture of baby formula containers, pacifiers, and bottles. Many companies have voluntarily stopped using it, but it's still in our food, soil, even the air.

The badness of BPA is epigenetic. The chemical adds undesired methyl groups to genes that affect the ways those genes behave, causing them to overrespond to estrogen. The research is not conclusive, but BPA has been linked with increased aggression, memory impairment, brain defects, obesity, and cancer. Hyperactive, hyperaggressive two-year-old girls are more likely to have moms who had high prenatal BPA levels. The hormone disrupter is linked to early puberty, diabetes, fertility problems, and cancers of the prostate and breast.

Wouldn't it be wonderful if we could eat something to protect our babies from BPA? In theory, we can. When scientists gave pregnant rats extra genistein (found in soy) and folate (found in leafy greens), the damage normally caused by BPA was blocked by healthy epigenetic effects. This is fascinating because it suggests a healthy diet could prevent or reverse damage caused by BPA exposure, among other epigenetic evils. Do women who shun leafy veggies and other nutritious foods during pregnancy have babies who are hit harder by BPA and other nasties like alcohol and nicotine? It's very possible.

These findings make me want to binge on vitamins, but there's a caveat. Researchers warn us not to stuff ourselves with folic acid and soy supplements to reverse potential BPA exposure. Too much genistein and folic acid during sensitive periods in fetal development may cause problems of their own. After all, they affect us the same way BPA does: by altering the behavior of genes. Until we know the correct dosage and timing, the best cure is abstinence: avoid soft plastics found in food packaging and the lining of cans. And eat a healthy diet—lots of healing greens.

CAN WE EAT TOO MUCH FOR TWO?

I've been eating like a truck driver. Yet here I am, daring to put my ungainly self on an uptown subway at rush hour. I'm squeezed between a shambling giant of a Rastafarian, a plumpish bald guy, and a petite blonde. As the train rattles along, the blonde glances over her shoulder at me but avoids eye contact. I understand she's agitated, but there's nothing I can do. *Believe me, lady,* I want to say; *I'm feeling it too.* We are lurching out of the Columbus Circle stop when the blonde whips around, exposing a set of tiny, foxy teeth, and barks, "Move it!"

"Move what?" I say sourly.

"The bag. Your bag in my back."

"That," I say, with impressive calm, "is my pregnant belly." Too bad I have too much padding for her to feel my baby kick.

I take the opportunity to reflect on my body, which has mush-roomed in the past five months. Truth is, I have gained seven more pounds than average for my starting body weight—and I'm continuing to fatten like a prize pig. To someone who has always been skinny, this is startling.

"You don't have to force yourself to eat," my obstetrician said tact-fully, allowing me dignity in the pretense that I gained too much weight deliberately.

Doctors and the pregnancy guides they write warn pregnant women that we do not have a license to be gluttonous. The parameters are well established: total weight gain for a woman starting with an average body mass index (between nineteen and twenty-six) is between 25 and 35 pounds: 2 to 5 pounds in the first trimester, 12 to 14 pounds in the second trimester, and 8 to 10 pounds in the third trimester. If we're carrying boys, we tend to gain slightly more than when we're car-rying girls (boys weigh an average of 3.5 ounces more at birth). Women carrying multiples should gain 10 to 20 pounds more than average. Heavy women should gain about 5 to 10 pounds less, and underweight women should gain 5 to 10 pounds more.

For most of us, this breaks down to about one hundred extra calories a day during the first trimester, the equivalent of a banana, and two hun-dred to three hundred extra calories a day in the second and third tri-mesters, the equivalent of two to three bananas. That's all. Yet more than 60 percent of us in the United States gain more than the recommended amount. Compared to pregnant women who gain 25 to 35 pounds, those who put on more than 53 pounds are more than twice as likely to have a heavy baby (over 8.8 pounds) who is at risk for obesity later in life.

The medical dangers of maternal overeating are well defined: preterm labor, gestational diabetes, hypertension, and an oversized (or sometimes undersized and underdeveloped) baby. We get increased pain of backache, varicose veins, a higher risk of C-section, and difficulty in taking off the weight later.

These warnings are practical and important to bear in mind. But what interests me—and truly motivates me to watch what I eat—is how my diet will affect my baby later on. I'm talking about lifelong consequences. For there's something interesting to learn here, not only about pregnancy but also about how we're wired.

A high-fat diet is obviously wrong—but that's what we usually crave. For ethical reasons, researchers can't directly study the effect of fatty diets on human fetuses by forcing women to overeat, but animal studies are revealing. At the Laboratory of Behavioral Neurobiology at Rockefeller University, neuroscientists fed one group of pregnant rats a diet with 50 percent fat, the equivalent of chocolate bars, potato chips, and pizza bagels, and another group a balanced, controlled diet with moderate fats. Diet was the only difference between the two groups of pregnant rats; they were genetically identical and fed the same chow until the equivalent of the second trimester of their pregnancies. In the last week of pregnancy, the researchers dissected brain tissue of some of the fetuses from both groups. The point was to see if fetal brains look any different when the mother eats a high-fat diet from when she doesn't.

The scientists were searching for anything out of the ordinary, and they found it. It turns out that the hypothalamus, the region of the brain that controls appetite and the release of hormones, was overgrown with neurons in rats whose moms overate while pregnant with them. Disturbingly, these neurons were sprouting much earlier than they should in a fetus's development. They were the type to give rise to obesity-inducing, appetite-stimulating molecules called orexigenic peptides. These peptides, once released into the bloodstream, bombard digestive systems with the ear-splitting signal "Eat more! More! You're *starving*!" They're like pushy Jewish grandmothers.

The researchers were curious to see if the early overgrowth of appetite-stimulating neurons in fetuses would mean fatter rats. After birth, both groups of rats were nursed by mothers on a moderate diet (the high-fat group was assigned to foster mothers), and then, after weaning, they lived on standard rat chow. Alas, it was too late for the offspring of mothers who were force-fed the equivalent of doughnuts and ice cream. Compared to the babies of moderate-eating moms, they ate more, weighed more, favored fattier foods, and reached puberty sooner. As the Rockefeller researchers see it, the babies were "programmed" that way in the womb.

This is another example of epigenetics—a mother's behavior affecting the way her baby's genes express themselves—in this case, producing appetite-stimulating neurons too soon and in alarming abundance. Mom's diet sends her fetus a signal about what the outside environment will be like. If she gorges on foie gras and potato chips, her baby is programmed to eat a lot of fatty foods too. Genes related to appetite are overexpressed, which results in eating more, and central nervous system mechanisms go haywire. Humans evolved in a world of food shortages, not excesses, and our bodies don't know how to compensate for gluttony. Sure, obesity is linked with diabetes, hypertension, cancer, cardiovascular disease, and age-related cognitive decline. But life is short. Why not eat cake?

I do mean short. But probably not sweet. Similar studies have found that rats—and human babies—exposed to fat-rich diets in utero are more likely to be obese, with a hankering for the sugariest, saltiest, and fattiest grub. They also have a higher risk of fatty livers, glucose intolerance, hypertension, and a reduction of immune system cells and immune system activity. Some studies have found that boys and girls born to obese moms are twice as likely to have attention deficit disorder and to be sadder and more fearful than their thinner peers. (Could America, with its soaring rates of maternal and child obesity, become a nation of neurotics and cowards?) The cause may be unrelated to prenatal diet, but it's possible that a mother's weight gain throws off her metabolism, which deprives the fetus's brain of nutrients or exposes it to the damaging chemicals stored in body fat.

That said, perhaps the most mind-blowing part of this is that epigenetic tags, which modify the way genes behave, can be passed down for at least two generations. If I blimp out throughout my pregnancy, I may damage not only my daughter's health but also add epigenetic tags to her egg cells that could someday be my grandchildren (the same goes for a son's sperm cells). To put it another way, my baby may be affected by what my mother ate during her pregnancy with me (she also craved doughnuts). It's enough to make me queasy.

The grandmother effect was studied at the University of Pennsylvania. Baby rats whose mothers ate junk food were born heavier, longer than average, and hungry as wolves. They were put on a normal diet, grew up, and had their own babies. And as it turned out, those babies—grandchildren of rats that pigged out in pregnancy—were, like their mother or father, born with longer-than-average bodies and a reduced sensitivity to insulin. Although the grandchildren were not overweight like their forebears, they were still predisposed to diabetes. Similar studies also found higher rates of breast cancer among females whose grandmothers ate high-fat diets while pregnant because they were born insulin resistant, which raises blood sugar, which in turn feeds cancer cells. (See "Why Grandma's (and Grandpa's) Diet Matters," below.)

I was curious about one more thing. Are girl fetuses more afflicted by their mother's diets than boy fetuses—as in what passes my lips goes to my daughter's hips? Or are sons harder hit? Curious about the same question, reproductive biologists at the University of Missouri divided pregnant rats into groups and fed each group one of three diets: high in fat, high in carbs, or moderate in fat. Halfway through the pregnancies, the researchers tested forty thousand genes in the placentas to determine whether activity varied with Mom's diet and the sex of the baby. And, as it happened, it did. Of the two thousand or so placental genes that expressed activity, those in females were much more active than those in males, and on high-fat diets more so than on low-fat diets. When genes activate in response to a high-fat diet, it's for a good reason: they're probably protecting the baby from harmful compounds.

This is fascinating because it suggests the placenta is something of a feminist—defending female fetuses more than male fetuses from the dangers of their mothers' high-fat diets. The best explanation is that male fetuses need more fat and calories to thrive. The male placenta may err on the side of caution—making sure plenty of fats eke through, but at the same time exposing him to potentially dangerous quantities of junk. This is not to say that female fetuses are unaffected. They are, but not as much as male fetuses. As in other animal studies, the researchers caution that we do not know if the outcome applies to humans yet, but there's reason to believe it does.

The bottom line is that I need to watch my big bottom. I'm not just eating for two; I'm eating for two generations. This research motivates me to cut back on cannoli, chips, and all the other carbs I crave. Epigenetic activity doesn't only suppress genes. It can also suppress appetite.

Why Grandma's (and Grandpa's) Diet Matters

In the winter of 1944–1945, the people of the Netherlands were profoundly worried about the future. Their country, whose motto is "I will endure," was falling apart. Nazi troops had blocked the distribution of supplies to the Allied-occupied West. By April, the Dutch subsisted on as few as five hundred kilocalories a day of bread or potatoes, about a quarter of the recommended daily allowance for a nonpregnant woman. Forty thousand women were pregnant at the time and suffered severe starvation. By May 1945, the country's distribution channels were restored, and life went on as usual. But the same can't be said of the babies who were born right after this crisis.

It turns out that the fetuses exposed to what has been referred to as the Dutch famine in the first trimester had less methylation, or suppression, of a gene called insulin-like growth factor 2 (IGF2) than their same-sex siblings who were born later. IGF2 allows a person to extract more energy from food by increasing insulin resistance. The children had a higher risk of health problems later in life, including obesity, diabetes, and clogged arteries.

This is classic epigenetics—tweaking how genes are expressed based on a faulty assumption about the future food supply. In a time of scarcity, increased insulin resistance is a virtue because it conserves energy. In a time of caloric abundance, however, thrifty genes that hoard every calorie of every supersized fast food meal are a liability. Let this be a warning to "pregorexics"—women who starve themselves in pregnancy to avoid weight gain. While they go skeletal, their kids may turn out obese and diabetic.

What's even more startling is the sleeper gene effect—the ghost of hunger past haunting grandchildren born sixty years later. What happened when the Dutch famine babies grew up and had families of their own? In a follow-up study, researchers found that the women had babies who were born at a lower birth weight and grew up to be shorter than average. It's as if their bodies were conserving, making them compact and efficient, in the expectation they'd be born into the famine that struck their mother when she was a fetus in her starving mother's womb.

Grandfathers too can influence a baby's genes, particularly if the men had an unusual childhood and the baby is male. Reviewing records of a nineteenth-century northern Swedish farming population that underwent feast-to-famine fluctuations in crop yields, researchers found that the nutritional status of boys between ages nine and twelve deeply affected future generations of their same-sex offspring. When a grandfather had a gluttonous season as a preteen, his sons' sons had a fourfold increase in the risk of diabetes and shortened life spans. But if a grandfather suffered from food scarcity at any point in his prepubescent years, those grandsons enjoyed longer life spans (even if his sons did not). Other studies have found that the age at which a man starts smoking affects the birth weight and early growth of his sons' sons. The earlier a man gets addicted to cigarettes, especially if it's before age eleven, the larger the body mass index (BMI) of his grandsons.

We don't know exactly how this "grandparent effect" happens, or why it sometimes skips a generation, but researchers believe environmental information is imprinted on sperm and eggs at the time of their formation. For boys, this happens in preadolescence. There may be a nutrition-sensing mechanism that influences certain genes in the sperm line, particularly those that control IGF2, which is active when passed on by men but not women. Cigarettes and other toxins may similarly affect the genes in preteens' sperm.

Among women, the prenatal period is when our diet can affect our children and grandchildren's genes. If we're pregnant with a girl, our decisions influence not only her but also the eggs within her that could someday become grandkids. We're like matryoshka, Russian nesting dolls—eggs within women within eggs.

We need brain more than belly food.

—American Proverb

DO PISCIVORES REALLY HAVE BRAINIER BABIES?

A lesson in eating the right foods during pregnancy comes from the lakes and wetlands of Africa's Rift Valley. Several hundred thousand years ago, something astonishing happened to hominids living in that watery world: they got smarter. Their cerebral cortex, a sheet of neural tissue in the forebrain, grew. The brain boost didn't happen all at once, but it was rapid by evolutionary standards. And as the cerebral cortex expanded, so did their memory, attention, perceptual awareness, thought, language, and all-around consciousness. Around two hundred thousand to five hundred thousand years ago, these lake dwellers had become so different from their ancestors and other hominids that anthropologists gave them a new name: *Homo sapiens,* meaning "wise man."

Bands of *Homo sapiens* migrated out of the Rift Valley forty thousand to a hundred thousand years ago and carry on today as you and me and our babies-to-be. And what can we thank for our ancestors' explosive brain growth? Their diet. All the fish they ate from the lakes and tide pools. For a long time, their diet was primarily fish and seafood because the forests shrunk and the climate had become arid.

This is the provocative theory introduced by geochemist Leigh Broadhurst and physiologists Stephen Cunnane and Michael Crawford. Studying conditions of the prehistoric Rift Valley, the trio speculates that our ancestors' diet of fish (and clams, mussels, frogs, and bird's eggs) boosted their intelligence enough to fish more often and more successfully, which in turn boosted their intelligence even further. Fish provides the essential nutrients for brain growth, which requires ten times more energy per pound than any other part of the body. Over tens of thousands of years, this virtuous cycle (along with other genetic and environmental factors) resulted in a shift toward bigger brains.

This theory of human evolution feeds the recent interest in prenatal benefits of fish and seafood. Brain development and function requires a flood of particular fats: docosahexaenoic acid (DHA) and its precursor, eicosapentaeonic acid (EPA), collectively known as omega-3 fatty acids. The foods richest in omega-3s are fish and shellfish. Looked at one way, our brains even resemble fish flesh: the omega-3 fat profile in fish is closer to that of the human brain than any other food known. Nuts and seed oils, egg yolks, and organ meats also contain omega-3 precursors. But marine life is the ideal fount for these fats.

In this spirit I have been taking two thousand milligrams of a salmon oil supplement daily. (Wild salmon, among the richest foods in omega-3s, contains about nine hundred milligrams in a three-ounce serving.) It's a chore. I retch on the translucent yellow capsules. Early on in pregnancy, I developed an aversion to their smell, so I make my husband uncap the container to avoid its fishy exhale. Peter doles the pills out to me as a nurse would for a recalcitrant patient, on a saucer or in a little cup. An implicit bribe is made: if I down them, I can have dessert.

This Herculean (or is it Neptunian?) effort is further inspired by studies that have found that fish oil consumption throughout pregnancy is associated with astonishing benefits for the baby-to-be, including improved language, visual motor skills, and problem-solving skills and less distractibility. Joseph Hibbeln, a biochemist and psychiatrist at the National Institutes of Health (NIH), led one well-regarded investigation. He and his colleagues followed nearly twelve thousand pregnant women in England who had filled out a food frequency questionnaire at thirty-two weeks gestation. The women were grouped into three categories: those who consumed no seafood, some seafood (up to 340 grams per week, the equivalent of two 6-ounce servings), and significant seafood (more than 340 grams per week). From six months to eight years after the women gave birth, their children were tested for verbal intelligence and behavior. After adjusting for factors such as social disadvantage, age, and maternal education, Hibbeln found that children whose moms ate less than two servings of fish and seafood weekly had an increased risk of falling in the lowest quartile for verbal IQ and a suboptimum outcome for social behavior, fine motor skills, and social development.

In another study, Harvard researcher Emily Oken and her colleagues recorded the weekly fish and seafood consumption of over three hundred mothers throughout their second trimester. Three years later, the researchers administered standardized cognitive intelligence and motor development tests to the children born of these mothers. After adjusting for multiple factors, the happy results were in: preschoolers whose moms ate fish more than twice a week showed improved performance on tests of language and visual motor skills compared to those whose moms never ate fish when pregnant. The difference was especially dramatic on tests of drawing, visual, spatial, and fine motor scores. The advantage was significant only when moms ate at least two weekly three- to five-ounce servings of fish or seafood during pregnancy. And the more often moms ate fish, the clearer the advantage.

Less optimistically, a recent study of more than seven hundred mother-baby pairs showed no effect of prenatal omega-3 supplements on the cognitive and language scores of the eighteen-month-olds. The pregnant women in that study took eight hundred milligrams of DHA supplement (although they didn't start pill popping until twenty-two weeks gestation, which may have missed a developmental window).

I want to know: How much does all this fish eating matter in the years upstream from now? Few studies have tracked mothers and their children for longer than the baby's infancy. In one, Hibbeln found that eight-year-olds whose moms took as little as one hundred to two hundred milligrams daily in a pill during pregnancy were significantly less likely to have suboptimal verbal IQ scores. In another, Norwegian four-year-olds benefited cognitively from their moms' prenatal consumption of cod liver oil, but they had no advantage at age seven, possibly because other factors such as social stimulation, education, and current diet are more influential by then. I think it's also interesting that in so many studies, children whose mothers ate fish flesh had better cognitive outcome than those whose moms took it as a supplement. Does the whole fish add up to more than the sum of its parts? For all these questions, time and further research will tell.

Although many studies show advantages of eating fish, they swim against the current of recent health advisory warnings. Two servings of fish per week, which was barely enough to produce a significant cognitive benefit to the children in the NIH and Harvard studies, is the limit recommended by the U.S. Environmental Protection Agency to expectant moms. The catch here is that while high levels of omega-3 may improve cognition, fish may also contain toxic levels of methylmercury (not in purified pills). Mercury and other toxins accumulate in brain fat just as they do in fish fat. A second-trimester ultrasound shows us the baby has a big bobble head on top of a shrimpy body, which is to say that all the fats go to the baby's brain at this stage, and so do all the fat-soluble toxins. Mercury can cause cognitive delays in children, including a lack of visual coordination and numbness.

As terrifying and real as the mercury threat is, researchers have tried to put it into perspective. Oken and her colleagues at Harvard did this by testing the mercury levels of their pregnancy volunteers. What they found is interesting. Among the expectant moms who ate fish less than two times weekly, those with the highest mercury levels did indeed have children who didn't perform well on the tests. Meanwhile, children whose moms ate fish more than twice a week during pregnancy had improved performance on the cognitive tests regardless of mercury levels. It may be that moms with the highest-scoring kids chose varieties of fish that are high in omega-3 but low in mercury, such as wild salmon, herring, and sardines.

Or they may have eaten the fish with foods that have damage-control properties. One intriguing Korean animal study found that garlic juice contains amino acids that suppress the uptake of methylmercury in the fetus's brain and reduce the amount of the toxin that crosses the placenta. Another experiment found that green tea, black tea, soy protein, oat bran, and wheat bran reduced mercury absorption in the gastrointestinal tract. Is it wisest to eat fish seasoned with garlic and soy, a bowl of bran on the side, and to wash it all down with a pot of tea?

The solution to the dilemma of whether to eat fish seems obvious to me. For the very real possibility that omega-3 fatty acids are beneficial to fetal development—as well as protein, vitamin E, vitamin D, and selenium—I eat low-mercury, low-toxin species of fish (wild salmon, sardines, anchovies), purified fish oil pills, or DHA-enriched eggs. Vegans may prefer to eat flaxseed and nut oils, which contain some omega-3s in a less bioavailable form. (For a complete list of high omega-3, low-mercury, sustainable fish, see Seafood Watch, http:// montereybayaquarium.org.)

If the connection between fish and human evolution is valid, then the implications are many. As the brains of our fish-eating ancestors got bigger, so did their skulls. And as their skulls got bigger, so did the female pelvis, from slim- to wide-hipped, to allow the passage of the big-brained baby through the canal. Looked at this way, the lakes and seas helped give womankind its curves. Neptune, god of the waters, should also be the god of beauty and wisdom.

Booty and Brains

Not long ago, two researchers, an epidemiologist and an anthropologist, wanted to know why men almost everywhere around the world prefer women with thinner waistlines than hips. William Lassek and Steven Gaulin were familiar with the prevailing theory that men prefer an hourglass shape because it is related to fertility, youth, and health. But they didn't believe that was the whole story. *Why*, they wanted to know, is it universally sexier to have padding on the hips and butt than the stomach or the back?

The answer, they discovered, is that not all fat is created equal. The waist and back blubber happens to be linked to brain inflammation, diabetes, and heart disease. But "sexy fat," the type that pads the hips, thighs, and butt (called gluteofemoral fat), is rich in brain-boosting omega-3 fatty acids. Throughout childhood and puberty, women store omega-3s in preparation for the third trimester of pregnancy and lactation, when the baby's brain will need it. The padding from the hips, butt, and thighs is the last fat that burns off when a woman loses weight or becomes malnourished. Only when having a baby does a woman's body willingly release this precious cache.

To see if there is a connection between booty and brains, Lassek and Gaulin crunched data of nineteen hundred mother-child pairs from a National Center for Health Statistics database, taking into account women's waist-to-hip ratios and their six- to sixteen-year-old children's scores on four cognitive tests. They came to a striking conclusion: regardless of weight, women who had a curvy low waist-to-hip ratio appeared to have smarter children, presumably because those babies benefited from their mom's ample gluteofemoral fat. Even after controlling for family income, race, and other factors, their scores on the tests were 2.7 percent higher on average.

That Mom's curves could put her kids ahead of the curve has some fascinating implications. The researchers contend that each cycle of pregnancy and lactation depletes the valuable stores of omega-3s (sagging butts, Moms?), which may help explain why back-to-back births could be a bad idea. Provocatively, depleted omega-3 stores may also explain why firstborns often have slightly higher IQs than their later-born siblings (who may have inherited fewer omega-3s), twins and multiples are at higher risk of compromised neurodevelopment compared to singletons, as are babies born to teenage moms who still need these fats for their own brain development.

All of us expecting girls have reason to pay special attention. Because girls store omega-3s in their hips and thighs, it's particularly important for them to get enough for brain development too. Lassek and Gaulin found that omega-3s accounted for twice as much improvement in girls' test scores as in boys.

Our omega-3-padded curves, of course, have other purposes as well. They keep our big-brained baby attached to us at the hip. The same may apply to the father.

WILL WHAT WE EAT NOW INFLUENCE
BABY'S TASTES LATER?

Eat fish with garlic sauce, and your amniotic fluid tastes garlicky. Eat apples, it gets appley. Anise makes it reek of licorice, and alcohol makes it boozy. Odorants from the food we eat flow in our bloodstream, trickle across the placenta, and wash up in the amniotic sea. Any strong flavor can make the migration. Women who eat curry the day they give birth deliver babies who smell like Indian takeout. Amniotic fluid is the real primordial soup—or marinade. All along, the baby swallows and wallows in the stuff.

No matter what flavors it, the swill is rich in nourishing proteins, carbohydrates, fats, hormones, and electrolytes. Like chicken soup, amniotic fluid strengthens respiratory, immune, and digestive systems. The broth is often yellowish: the fetus pees in it, drinks, pees again, and drinks again. It's salty. Bobbing in it are flecks of vernix, the waxy coating that flakes off the fetus. The fluid also contains cells shed from the baby's skin, sac, and digestive tract, which is why several tablespoons may be sampled for amniocentesis, a test for chromosomal abnormalities. Amniotic fluid is reabsorbed and replaced daily at the beginning of the second trimester, every few hours by the end of the second trimester, and even more frequently in the third trimester.

I am now in the middle of my second trimester, which is when my baby should gulp down more than two soda cans worth of amniotic fluid daily. This is also the time when a fetus starts to remember and fancy familiar flavors. Taste buds form as early as six weeks after conception, but the baby has to wait months for her olfactory system to go online and the mucus plugs in her nose to dissolve (if you've ever had a cold you know you can't taste well without smell). Now, at last, the fetus's olfactory cortex, where odors are processed, knows to send signals to a way station (the thalamus), where odors are identified. The odor signals are then sorted out (in the orbitofrontal cortex) and sent to three crucial regions of the brain: where smell is associated with emotion (amygdala),where appetite is regulated (hypothalamus), and where the memory of the odor is filed (hippocampus). Odor processing motivates the baby to associate amniotic odors and tastes with Mom and to prefer those smells and flavors after birth.

As for the amniotic fluid my own fetus is marinating in, I suspect it tastes rooty. Carrots are one of my obsessions in pregnancy. I like them raw, dirty, and unpeeled. I eat so many carrots that I can see why people say I have a basketball in my belly—or is it a rabbit? I wonder: Am I programming my baby to be a carrot lover?

This question happens to be addressed in a fascinating flavor-learning experiment at the Monell Chemical Senses Center in Pennsylvania. Researchers recruited pregnant women and divided them into two groups: one drank about 1¼ cups of carrot juice four days weekly for the last three weeks of their third trimester, and the other drank water. The mothers went on to give birth, and by the time the babies were about six months old, the mother-and-baby pairs returned to the lab at Monell. Curious to see if infants who had been exposed to carrot flavors in utero would respond differently from others, the researchers flavored some cereal with carrot juice and recorded and rated the infants' reactions.

Sure enough, babies who had been marinating in carrot-flavored amniotic fluid expressed more enthusiasm and less disgust when fed carrot-flavored cereal than babies not prenatally exposed to carrot flavors or exposed only through breast milk. They also fed longer. It's safe to say that prenatal exposure to flavors makes babies respond favorably to them. "Significant traces of this may remain as children become adults and pass on their food habits to the next generation, often via amniotic fluid and breast-milk associated cues," write the researchers—unless, or until, they're tugged in other directions by peers, advertising, and food availability.

It's one thing if my baby likes carrots. I'd love her to have a lifelong love of vegetables. But how do I feel about her brain literally being shaped by all the foods I've been feasting on, including the flaky, crispy, sweet, and syrupy stuff? Researchers at the University of Colorado at Denver fed one group of pregnant mice a mint-flavored diet and the other a bland diet. After they were weaned, the baby mice whose moms ate the flavored diet had significantly larger glomeruli, a region of the brain involved in detecting and discriminating among specific odors. These mice also strongly preferred the same minty food their moms had been eating in pregnancy because their glomeruli made them more sensitive to minty flavors.

That human babies too could be hardwired to detect and favor flavors they experienced in utero has an evolutionary basis. In the past, prenatal flavor learning would have been beneficial because flavors tasted in the womb would represent foods that were safe and available in the environment. From the fetus's perspective, whatever tastes he encounters in utero are good. After birth, he'll taste these flavors again in breast milk. A baby who is sensitive to the tastes and odors in his mother's diet is more likely to be better at sourcing those foods and distinguishing them from similar but less appropriate ones.

Fetal flavor learning works in my baby's favor when it applies to carrots but not to carrot cake—and the many other sweet carbs I crave and consume in pregnancy. It's possible, even likely, that fetuses are born to crave the taste of sweets if they've been exposed to a constant drip of sugars in utero (diabetic women have high sugar levels in their amniotic fluid and are more likely to have babies who become diabetic). I wonder if this is one reason that so many of my friends born in China, where desserts are less desired, have less of a sugar addiction than I do.

An addiction to the other white stuff, salt, may also be traced back to the womb. Salt flavor learning has less to do with what mothers consume than what we can't keep down. Oddly enough, women who reported moderate to severe vomiting episodes gave birth to salt lovers. At sixteen weeks old, their infants drank saltier solutions and made sweeter expressions when drinking it compared with their peers whose moms weren't so sick. Even when their kids are college students, as another study by the same group found, they're far more likely to crave the salty stuff. One explanation is that vomiting mothers lose a lot of fluids and minerals, which permanently programs their fetuses to compensate by craving salt. High salt intake increases fluid retention, a protection against dehydration.

Sugar and salt cravings are worrisome enough. Even more nagging is the idea that moms who drink alcohol or smoke when pregnant are programming their children to crave alcohol and nicotine. When rats in the equivalent of the third trimester were forced to drink a moderate dose of ethanol, the equivalent of two beers a week, their offspring drank more ethanol than those whose mothers did not consume it in pregnancy. Similarly, women who smoke while pregnant are more likely to have teenage daughters who smoke, regardless of their upbringing. Although human fetuses, seen in ultrasound, clench their doll-like lips when their amniotic fluid is laced with bitter-tasting alcohol or nicotine, they're still bathing in the stuff and their brains may well habituate—leading to an appetite for, or even an addiction to, booze and cigarettes later in life.

So fascinated am I by fetal flavor learning that I have become more mindful of my dessert consumption. How much can I safely sweeten the amniotic fluid? Judging from my baby's chocolate-induced kicks (and, I presume, generous gulps), I have a feeling she's got a sweet tooth long before she teethes.

> The truth lies in between the 1st and the 40th drink.
>
> —Tori Amos

IS A LITTLE TIPPLE REALLY SO TERRIBLE?

The winter holidays have just passed, and I am reading a discussion in an online fertility forum. A woman who goes by the name Gonnabemom is in a fresh panic. On New Year's Eve she made lentil soup and added two cups of white wine. She ate the soup and felt tipsy. The following day she took a pregnancy test, which came out positive. Her emoticon is a mouth frozen in midscream, bulging eyes flashing in fear.

The community rushes in to save Gonnabemom. "No worries," writes NYCbabe, "the alcohol evaporates when you cook it." Nosexin-

Texas chimes in to say her DH (dear husband) made a superb cognac peppercorn reduction steak sauce the other night, which she relished with a half-pint on the side. A pregnant poster who calls herself Prego-ragu wants to quibble with the first responder. "Actually," Pregoragu writes, "alcohol doesn't go away when cooking; there will still be some left. But don't worry, hun!"

Worry, however, seems to what most expectant mothers do, whether they're concerned about a few cocktails downed at a bache-lorette party during the two-week wait or the occasional nightcap in the insomniac months that follow. Much of the confusion comes from the conflicting information put out by the medical community. There is no consensus about how much alcohol may be safely consumed dur-ing pregnancy.

Some studies, usually European in origin, lift the spirits. There's no apparent reason to avoid light drinking, according to one at the Uni-versity of London of nearly 12,500 mothers and their three-year-old children. Reading the results, I contemplate swapping my wheatgrass for a tequila shot. It turned out that children whose mothers had one or two alcoholic drinks per week or per occasion during pregnancy had fewer behavioral and cognitive problems by age three than the tots of teetotalers. Specifically, the sons of light drinkers were less hyperac-tive and had higher test scores. The daughters had fewer emotional and peer problems.

The authors of the study did not speculate on why drinking seemed to have this surprising effect but urged readers to interpret the results with caution. The effect may be due to something unrelated. Are light drinkers more affluent and better educated? Are their stomachs full of healthier food? They likely had their drinks with meals. Perhaps they ate high-omega-3 fish with their chardonnay, which in some studies is linked with fewer behavioral problems in children. Maybe the anti-oxidants in wine offset the dangers of the alcohol. Or is there some-thing about pregnant light drinkers that makes them better mothers? Maybe they're less anxious.

True neurotics will dig deeper. Problem is, we're hard pressed to find any evidence that light drinking (a weekly or occasional drink) during pregnancy, especially after the first trimester, has any profound effects on the developing fetus. Moderate drinking—between one and three drinks daily—is associated with somewhat adverse birth outcomes; babies may be born underweight or preterm and may have shorter attention spans as toddlers. Most obvious, of course, is the impact of binge drinking—downing four or more drinks a day. Heavy drinking is associated with hyperactivity, inattention, and other academic difficulties in kids. At the extreme are babies born with fetal alcohol syndrome (FAS), a disability that causes abnormal facial features and behavioral problems, mental retardation, poor coordination, heart defects, and growth deficiencies. (Visible abnormalities can be seen in a second-trimester ultrasound.) How many drinks did the mother of a FAS baby drink? Probably more than five daily, or several binges weekly, but there are no clear-cut thresholds.

Because the alcohol safety limits are so ambiguous and the stakes are so high, many pregnant women become pious teetotalers. This isn't true for all nationalities: the French are more likely to shun salads (for fear of Listeria) than the occasional glass of chardonnay. The Irish might take a swig of stout for good bloom. Australian midwives might prescribe a middy (a half-pint). Pregnant Brits traditionally gulp Guinness for "the iron."

Iron, incidentally, is key to an intriguing new view on why it's so difficult and controversial to determine alcohol's safety in pregnancy: some races have a higher tolerance than others. According to population geneticist Tom Cochran, people whose ancestors lived in agricultural communities in Europe and the Middle East between six thousand and ten thousand years ago developed a genetic tweak (of the alcohol dehydrogenase genes, ADH1 and ADH2) that enabled them to better metabolize alcohol, which, back then, was made from fermented barley rich in iron and other nutrients. Those who could metabolize alcohol efficiently and extract its blood-building benefits had a nutritional edge over those who could not. Iron-deficiency anemia is a pregnancy risk, so women with the gene variant were better able to nourish themselves and their babies by quaffing the nutrient-rich swill. These women were also healthier because alcohol was safer to drink than water, which was often contaminated.

Studies have found that fetuses that inherit more efficient alcohol-metabolizing genes—in particular, a certain variant of the ADH1B gene—are better at protecting themselves against fetal alcohol syndrome. They may also be less exposed to the toxins in alcohol thanks to their mother's ability to metabolize it. It's a shame such gene variants did not make deep inroads into Asia, Africa, or the New World, which may help explain why FAS shows up almost thirty times more often in many of those populations than in babies of European or Middle Eastern ancestry, even if both groups consume the same quantity of alcohol.

Our diets also matter when it comes to how blitzed a fetus is by her mother's prenatal alcohol intake. One recent study found that among women who drank in pregnancy, those who took multivitamin supplements were no more likely to have a miscarriage than women who abstained (an epigenetic mechanism in the nutrients may reverse the damage to genes). If you're well nourished *and* have the genes to metabolize alcohol better, you have a royal flush: your unborn baby is less likely to be afflicted by alcohol exposure.

But even people with the best alcohol-processing genes and diets can't count on them to guarantee a risk-free pregnancy. After how many steins do we cross the line? No one knows. There are too many variables. Perhaps someday we'll be able to compute our individual risk exposures. In the meantime, we all continue to follow the old drinking dictum: when the wine goes in, you never know what comes out.

4

PUDGY HUBBIES, MAN BOOBS, AND A THEORY ABOUT THE DADDY GENE

A Daddyology

Well into my second trimester, I'm in a café eating a lunch of baguette and cheese with a mug of hot chocolate. My dining companion is my friend Chris, a childless bachelor in his forties. We're discussing my new favorite topic: babies. Chris chuckles cynically.

"I always see men walking around with infants strapped to their chests in kangaroo pouches," he remarks. "I could never do that."

"Perhaps you would if you were expecting a baby," I say gently.

Chris smirks. "I don't know if you know it," he says, "but those guys go out of their way to avoid eye contact with other men. They know how they look to us!" He gives a little laugh and shakes his head. "But isn't it interesting," he says with a wistful sigh, "how attracted women are to wimps? I went out with my friend and his baby daughter. Women were all over him! Babies are babe bait!" He chomps into his ham sandwich.

I take on the mantle of informing Chris that women seek out men with paternal qualities. They're sexy. Look at Brad Pitt. I tell him to take a look at his competition on online dating sites. Isn't he aware that men post photos of themselves frolicking with kids for a reason?

Chris doesn't seem to hear me. I watch him gaze pensively out the window at the people on the street. "Another one," he says with a mannish snort. "They're bloody everywhere." I follow his eyes and see a man walking by wearing a newborn in a Baby Bjorn front pack carrier.

No one knows if early male *Homo sapiens* carried babies in slings, but it's likely that there was a great deal of paternal care among our ancestors. Not long ago, many scholars would've disagreed. The iron-clad rule was that mothers nurture and fathers provide. But that stereotype is getting overturned in favor of one that suggests that ancestral fathers—along with grandparents, older siblings, aunts, uncles, cousins, and friends of both genders—were also active nurturers. How else but with a great deal of care and support could our big-brained species have such long childhoods in a world of scares and scarcity? In about 10 percent of mammalian species, fathers are involved in child rearing. Human beings are one of them.

Sarah Blaffer Hrdy, an anthropologist and professor emeritus at the University of California-Davis, and a mother, promotes the "it takes a village to raise a child" view of history. Look at evolutionary patterns, she'd say. Traveling bands of hominids must've relied on fathers (among other blood relatives) for child care, for all the work couldn't have fallen on mothers' shoulders only. Look at the way men in contemporary preindustrial cultures raise their kids. The Aka in the western Congo are within eyeshot of their babies 90 percent of the time. They may be the world's most affectionate fathers.

Even the Chrises of the world might change the way they act when they themselves become fathers. Paternal behavior is a characteristic of our species. Proof of this, say Hrdy and her colleagues, is in men's hormones and behavior, since they shift in the presence of pregnant mates and newborn babies. Many become less aggressive, more emotionally sensitive, and more nurturing. This wouldn't happen if the capacity for nurturing weren't in men's genes.

Yes, human males evolved to behave dadly. This shouldn't be a radical statement. But we have been socialized—or brainwashed—to think manliness is incompatible with baby and child care. Pregnancy is not a prerequisite for parenting. As more women gain economic power and become the breadwinners in a family, we see more Mister Moms staying home with the kids. It does take a village to raise a child, and the most desirable villagers may be the ones willing to bear the Baby Bjorns.

His breasts are full of milk, and his bones are
moistened with marrow.

—Job 21:24

WHY IS HUBBY GETTING SICK AND CHUBBY?

"We're pregnant!" I announce to a friend with three grown children.

"No, dear," she gently chides. "Unless you're using the royal we, which you're entitled to do in your state, it is only *you* that's pregnant. It's all you, darling! Your husband can sympathize, but he'll never understand."

"You have to see him," I say. Peter acts as if he's pregnant too. We are a frumpy Tweedle-Dum and Tweedle-Dee, bumbling along, bellies thrust forward, in sweats and the spongiest of sneakers. We both have banana cake cravings and flatulence. We have a low threshold for foul odors. Walking on the city streets behind an overperfumed old lady, we'll look at each other, wrinkle our noses, and dash in front of her to avoid the noxious slipstream. We blanch at the mere whiff of sewer gas. We can determine, by subtle shifts in sourness, whether the orange juice we are drinking was squeezed in the morning or the afternoon. We have zero tolerance for smokers.

Poor Peter. None of this is manly.

Nine of every ten expectant fathers are like my husband, afflicted with at least one of their partner's symptoms: mood swings, nausea, fatigue, food cravings, odor aversions, and bouts of bloat. The attacks strike in the first trimester, wane in the second trimester, and return with a vengeance in the third trimester. Nearly half of these men gain thirty pounds or more. Some grow firm little man boobs.

This is no new age, metrosexual, sensitive-guy phenomenon. Even in antiquity, men experienced pregnancy symptoms and ritually partook in their mate's pregnancy. This included abstinence from some foods, especially animal flesh that may contain toxins; restriction from hard labor; and pampering by kinfolk. Marco Polo, who observed male pregnancy symptoms among the thirteenth-century Chinese, considered them to be a form of sympathetic suffering. Travelers in the West Indies reported how fierce tribesmen retired to the men's clubhouse when their wives went into labor. According to *The Custom of Couvade*, an anthropological survey in the early 1900s by Warren Dawson, men would writhe in pain, "even simulating by groans and contortions the pains of labour and sometimes even dressing in a wife's clothes." In some cultures, Dawson described, men lay wasted in bed after the baby was born and were even waited on by the newborn's mother. The symptoms may be heartfelt, biological, or culturally enforced—or all three, it's not clear—but the outcome is the same. "You would swear it was he who had the child," wrote an astonished eighteenth-century Jesuit missionary in South America.

Observing these ruffled men, psychologists termed male pregnancy symptoms *couvade*, from the French *couver*, meaning "to incubate" or "to brood on the nest." Male birds in couvade assume the role and appearance of mother birds by warming the eggs laid by females. Couvade takes on various meanings when applied to men. Is it, as presumed in the Freudian nineteenth century, a form of fetus envy—the male analogue to penis envy—a longing to be the one with the bulge in front? Among the alternative explanations are a literal envy of the fetus (the plumpish man wishes to be a baby again), pseudo-sibling rivalry (competition with the fetus for the woman's attention), or anxiety about fatherhood (a desire to escape into his own body fat).

Twenty-first-century science has revealed that couvade isn't so cuckoo. It's an evolutionary adaptation, in a biparental species, to prepare the father for the birth of a baby. Men come by it naturally. So do other male mammals that share parenting duty, including marmosets, hamsters, voles, and tamarins, which gain weight even faster than their pregnant mates do. Primate males who help raise their young consume significantly more food when their mates are pregnant, gaining as much as an additional 20 percent of their body weight. From an evolutionary perspective, paternal pudge is an energy reserve. It's nature's way of preparing males for parenting, one of life's most calorie-intensive activities.

To explain on a biological level how this happens to expectant fathers, psychologist Anne Storey and zoologist Katherine Wynne-Edwards turned to prolactin. A Buddhist of hormones, prolactin slows everything down to a more meditative in-the-moment pace. Under its influence, fats and sugars are metabolized less effectively, resulting in weight gain. We get Buddha bellies, bums, and boobs. Fat tends to stick around.

So, incidentally, do friends and family, because prolactin (and progesterone, another pregnancy hormone) is associated with bonding and caregiving behaviors. Prolactin arouses sweet and tender feelings by raising pleasure hormones known as opioids. Also heightened are the senses, particularly smell, because prolactin and estrogen stimulate new neurons that migrate to the brain's olfactory bulb. In men, prolactin lowers the libido and shrivels the erection. In women, prolactin makes the milk flow. High on prolactin, we're dreamy, heavy, soft, abundant, less sexual, and emotionally and sensorially astute. We're about to enter our new incarnation as parents.

It would be enlightening, Storey and Wynne-Edwards thought, to test the prolactin levels of expectant fathers, so they recruited thirty-one couples, mostly first-time parents, from a parenting class at a Canadian hospital. Some of the couples had recently had a baby, and others were expecting theirs to arrive soon. The men and women agreed to have their blood drawn before and after they held their newborn or, in the case of the prenatal couples, before and after holding a soft-bodied doll swaddled in a receiving blanket. Both groups watched a video of a newborn struggling to breast-feed for the first time. They also completed a checklist of pregnancy symptoms about themselves and their partners. Did they have weight gain, nausea, increase in appetite, and emotional changes? If their baby cried, did they feel irritated, excited, anxious, or concerned? How stressed out did they feel?

As Storey and Wynne-Edwards predicted, many expectant dads had prolactin concentrations that mirrored the soaring levels of their partners late in pregnancy. The more emotionally in tune a couple was around the time of birth, the more likely the male partner's prolactin level was high (although the woman's was always higher). These couples were like two birds in synchronized flight. And the more elevated a man's prolactin levels were, the more extreme was his couvade—that is, the more fat reserves he gained, the more crippling the nausea and the stronger his food and odor aversions.

There's an important lesson here for expectant moms: rejoice if your partner has some of your symptoms. The guys with the nausea, weight gain, and other high-prolactin side effects were more attached to their newborns and more emotionally responsive than the low-prolactin papas. Men who had been in couvade were also the flappable type, clucking and cooing concernedly when their newborns cried. Their levels of the stress hormone cortisol were higher.

Their testosterone levels also dropped by about 30 percent on average—arguably making them less competitive and more compassionate. Nosy me wonders: Did this hormonal downswing happen to Tiger Woods? The golf pro was on a losing streak around the time his wife gave birth to their son. Yet in interviews, Tiger seemed content. He gushed about his domestic life and posed for portraits with wife, kids, and labradoodle. Was his prolactin high and testosterone low, as it is in so many new fathers? All we know is that Tiger got his game back later (in more ways than one). Presumably his prolactin levels plummeted and his testosterone rebounded. The hormonal honeymoon normally lasts only four to seven weeks after the birth.

Every time a man is expecting a new baby, the hormonal effect may rebound. In a follow-up study, Storey and her colleagues found that second-time dads have even stronger prolactin surges than first-time fathers. Experience primes men to hear and smell babies better and respond to them more compassionately. Their bodies must hold a memory of babies past.

Surprisingly, the hormonal tide ebbs and flows with a father's contact with his children: prolactin drops when he has been around them for a long time, and it soars when he returns after he has been away. This may sound counterintuitive, but there's an evolutionary explanation. If fathers' prolactin levels were always high, nurturing might conflict with other responsibilities. Fathers have traditionally been providers too, and boomeranging hormone levels might help them strike the right balance between bonding with the baby and bringing home the bacon.

There is a powerful lesson in the story of prolactin. Men evolved to be fatherly, and the hormone is evidence that paternal behavior runs in their blood. Although mothers can mother without them, fathers have always been a vital part of child raising.

In this spirit, I'm going to keep saying it: *We're* pregnant. *We're* uncomfortable and excited, sleeping too little and eating too much. Of course I'm bearing the brunt of the burden. But I'm happy to give my husband some credit for suffering alongside me. As long as he isn't being a big baby about it.

> [He] had seen men hug their wives, the way they'd fit their chin down over the woman's shoulder and there would be this smile, a particular young-seeming grin with closed eyes—always made him think—bliss.
>
> —A. L. Kennedy, author

DOES OUR SCENT AFFECT OUR PARTNER SUBCONSCIOUSLY?

Not long ago, twenty pregnant women in their thirties committed themselves to a strange yet completely harmless experiment. Three times—once in their first trimester, again in month nine, and then six months after giving birth—they agreed to stick patches on their armpits and nipples for at least twenty-four hours. A day or so before applying the patches, the women were forbidden to wear deodorant or certain fragrances, and they had to restrict their diets to bland foods. They were strictly required to report any emotional disturbance while wearing the patch. This was very important because anger or stress could throw off the experiment.

The researchers overseeing the study—an organic chemist, an evo-lutionary biologist, an obstetrician, and an epidemiologist, all of them Italian—were curious about what they'd find in the pregnant women's sweat. They were looking for something very specific: chemical com-pounds associated with pregnancy. An interesting finding would be one in which certain chemicals are present in the sweat and secretions of pregnant women but not in nonpregnant, nonlactating women. In essence, the scientists were hunting for pregnancy-specific phero-mones.

Cheapened by fragrances and soaps that promise to brainwash sex partners, pheromones get a bad rap. They shouldn't. They're simply air-borne chemical signals (chemosignals) that trigger predictable behav-iors in other members of the same species. Dogs use pheromones in their urine to ward off other canines. Aphids use alarm pheromones to mount a unified defense against an attacker. Pigs and cats and hun-dreds of other species use sex pheromones to put others in the mood for mating.

Given that the rest of the animal kingdom uses pheromones with impunity, it seems very likely that we do too. These chemical signals are found in our sweat and other bodily essences. When others smell us, our pheromones hit their hypothalamuses, where hormones are released. Pheromones may be behind some everyday weirdness: syn-chronized menstrual periods among women in close quarters, acci-dental pregnancies due to early ovulation, one-night stands, the mysterious feelings of attraction to or repulsion from certain people, mood contagion, and so on. Unconsciously, we may emit chemical sig-nals that advertise to others that we are, say, fearful, fertile, horny, or genetically compatible. Or pregnant.

It was a eureka moment for the Italian scientists when the pits—and nipples and areolae—of the pregnant volunteers contained what could well be pheromonal gold. The lab detected five chemical com-pounds found only in the sweat of expectant moms and not in women who weren't pregnant: oxybis octane, which is used to make perfume; hexyl-cinnamic aldehyde, a major chemical in calming chamomile; isocurcumenol, involved in modulating neurotransmission; and dodecanol and isoproprylmyristat, sex pheromones that insects and crustaceans use to attract mates. By the third trimester, women were pumping out a payload of these chemicals, which have a subtle but distinctive smell. Six months after the women gave birth, regardless of whether they were breast-feeding, only one chemical (isoproprylmyristat) was detectable in a significant quantity.

Why do we put out these chemical signals only when we're pregnant? One explanation is that we'll need them to communicate with the newborn. A homing signal of sorts, they might help infants identify us and find our nipples. But an even more provocative theory bubbles up: Might body odor during pregnancy also target the father of the baby?

To my surprise, Stephen Vaglio, one of the Italian researchers, says yes when I ask him the question. "The chemosignals that we found *might even help couples bond* toward the end of pregnancy [italics mine]." Four of the chemicals we put out are more abundant during pregnancy than afterward, which suggests they're meant to play a role before the baby's arrival. The theory goes that when a man kisses his pregnant wife, has sex with her, sniffs her sweat, or spends a great deal of time in her presence, he picks up on these chemical signals. Once inhaled, the pheromones are processed in his hypothalamus, the part of the brain that triggers the production of hormones. Could these chemicals coax our male partners to be hormonally in sync with us, especially in the third trimester when we need them to be sensitive and responsive? It's an intriguing theory. Maybe this explains why expectant fathers produce more prolactin and less testosterone, a hormonal shift that makes them better dads and mates. Our odor is the trigger.

A devious delight fills me. Do I leave pregnancy pheromones in my wake? Do they trail behind me like fairy dust or radiate around me like an invisible aura? Maybe I leave them like love charms on doorknobs, bath towels, and pillowcases. Does my husband inhale these chemosignals when he kisses me? Are they messing with his mind? And what about all the taxicabs and diners and public restrooms I've been in?

Of course, whether chemicals in the sweat of pregnant and recently pregnant women have a bonding effect is an open question. Promisingly, an unrelated study found that the sweat of breast-feeding women acts as an aphrodisiac of sorts (see "Is Our Sweat Sexy?" page 219). But everything we know about pheromones suggests they are context dependent; just inhaling them probably doesn't do much. Only an expectant father who is both emotionally close to his partner and lives with her may respond to the chemical trigger. It seems to me that a good deal of exposure and emotional priming is required. After all, male obstetricians and nurses don't necessarily feel a pull. Or do they?

The point of this research is that it supports the evidence that people, like most other animals, emit chemical signals that affect the behavior of others. We require support during pregnancy and after the birth, so it'd make biological sense for our mates to chemically bond with us. This is power. We don't need to lift a finger as long as we raise our arms.

> A primary role of the father is to teach the baby that love and comfort are not necessarily associated with food.
>
> —Anonymous

CAN MEN BREAST-FEED?

Here's the next puzzle: If the hormone prolactin triggers lactation and men get prolactin surges when they hang around their pregnant wives and newborns, then why can't men breast-feed? After all, guys have nipples. We've all seen man boobs. What's to stop those boobs from nursing?

Although my husband looks forward to feeding our baby expressed milk from a bottle, he doesn't appear to appreciate this line of inquiry. Lactation is where Peter draws the line. He folds his arms against his chest while I rattle on about papa's milk.

"No," he says, shaking his head.

"Yes, we'll both breast-feed!" I happily insist. There is a beloved passage in Charles Darwin's *The Descent of Man* that goes, "In man and some other male mammals . . . [the mammary glands] have been known occasionally to become so well developed during maturity as to yield a fair supply of milk." In the 1896 *Compendium of Anomalies and Curiosities in Medicine*, there are a few cases, perhaps apocryphal, of village widowers and male missionaries in Brazil who've lactated to save newborns in the same miraculous way that grandmothers have lifted trucks off toddlers. Not long ago, a thirty-eight-year-old Sri Lankan man, a Mr. B. Wijeratne from Walapanee, allegedly nursed his two daughters after his wife died in childbirth. But how—and why don't I hear about this more often?

The good news is that there's no insurmountable barrier to male lactation. Not even pregnancy is a prerequisite. Virgin females and postsurgical patients can spontaneously express milk (after inadvertent stimulation of the nipple nerves), and so do some women who take birth control pills (from hormones). On farms, male goats have expressed milk from what appear to be swollen teats. The male Dayak fruit bat of Malaysia occasionally produces about five micrograms of milk, although whether it suckles its young is debatable. The cups of males may not runneth over, but they are not empty.

The lack of lactation among male *Homo sapiens* is not the result of a hardware limitation. The nipples of boys and girls are essentially identical, with the same ductwork and lobes, until puberty, when testosterone suppresses further development in males and estrogen coaxes fatty tissue growth in females. But one doesn't need a big bosom to express milk. For instance, thousands of men who had been prisoners in World War II got wet spots on the front of their shirts soon after they were released and ate their first hearty meal. The source was a thin blue-white gruel secreted from their nipples. Breast milk.

Lactation can't happen without prolactin, a hormone that controls milk production. Prolactin is triggered by the hypothalamus, the hormonal command-and-control center in the brain, which sends a signal to the pituitary gland, saying "Milk! Now!," and prolactin is sent forth in tidal waves. Nipple stimulation triggers a chain reaction: the more suckling, the more prolactin, the more milk.

For my husband to breast-feed our baby, he'd need to produce a lot of prolactin. This happens naturally when there are problems with the hypothalamus, the pituitary gland, the liver, or the testicles—which normally keep prolactin in check. For instance. the pituitary gland could overproduce prolactin and other hormones, resulting in nipple leakage. In the case of the lactating prisoners, eating triggered their hypothalamus to produce extra prolactin, which went unchecked because their livers were damaged and testosterone levels were low. The same happens to male alcoholics.

To Peter's relief, the prolactin rush men get around their pregnant wives and babies is nowhere near the amount required for lactation. While he and other men are happy to leave it at that, case closed, I am not so ready. I remind him that men not only have nipples but hormone producers and receptors, enzymatic pathways, and other breast-feeding physiology. In other words, he's got the right hardware. With the right software running—that is, more prolactin—I can imagine male lactation as a hormonal response that comes as casually as an adrenaline surge.

So why haven't we seen Male Lactation 1.0 yet? Of the four thousand-plus species of mammals, none lactate regularly. The answer from evolutionary biologists is cold and crass: it just doesn't pay—never has and never will. While fathers can be nurturing and individual men can be exceptionally paternal, it's just not worth it for them to breast-feed. Historically, fathers have had a lot less invested in each child's survival than have mothers. They don't put as much time, energy, and effort into pregnancy as we do, to the exclusion of other reproductive opportunities. A man also has to deal with paternal uncertainty—he can't be sure the baby is his own biological issue. Why undergo feminization to save what might be another man's child? To the extent that hormones influence behavior, a prolonged prolactin surge could make a man less aggressive, less territorial, less competitive, and less attractive to other women. His sex drive would screech to a halt.

According to biologists Thomas Kunz and David Hosken, several conditions must be met for it to make evolutionary sense for men to naturally breast-feed their young. Two of these are food unreliability and obligatory biparental care. These pressures have never been on men, they claim, because babies have other fallbacks when moms are absent, including wet-nurses, and now, formula (although there is still no perfect substitute for breast milk; see "What's Living in Our Milk (and Why)?" page 206).

From a biological standpoint, for men to evolve the ability to breast-feed, we'd also need to live in a society in which guys can freely feminize. This can happen only when there's less rivalry among males and fewer external threats—when it's more advantageous for men to cuddle than to compete. This is not the environment humans evolved in, and it is not necessarily the world we live in now.

But it makes one wonder. While it's unlikely that all these conditions would be met under normal circumstances, lactation can happen by man-made means. A course of hormonal priming and prolactin-boosting drugs followed by suckling stimulation can help anyone breast-feed—often adoptive moms and gay fathers—even if they may not be able to produce enough milk to sustain a baby without supplements. Why don't all dads think about doing this? Perhaps breast feeding should be a viable option for men in the twenty-first century, especially for the many fathers who feel alienated from their newborns because they can't nurse them. With modern medicine, any man can become a milkman.

My insufferable enthusiasm makes Peter even more resistant.

"I don't think this one is going to fly, sweetie."

But who knows what fatherhood will be like by the time my daughter or her children become parents? I imagine that if culture evolves and becomes more civilized and cooperative, the meaning of manhood may mutate. Guys may feel more comfortable assuming traditionally feminine roles, including being the primary caregivers to children. Women may choose men for their ability to nurture. This is happening already. For men, breast feeding is just one braver step in that direction.

But will they be man enough to do it?

Fatherly Faces

Over the past few years, a number of psychologists have begun to look at how facial features reveal qualities about a person. The bulk of these studies involve recruiting a group of unsuspecting volunteers and asking them to make snap judgments about strangers who have been pretested for certain traits. This is exactly what psychologist James Roney and his team of University of California at Santa Barbara researchers did to determine if women could glean men's interest in babies, among other qualities, just by looking at their mug shots. The researchers took photos of thirty-nine heterosexual men ages eighteen to thirty-three, tested their saliva for the hormone testosterone, and assessed their interest in infants. Then they asked thirty women to rate each man's headshot for qualities such as physical attractiveness and masculinity, fondness for infants and children, and suitability in a short- or long-term relationship.

Although they had only a still photograph to go by, nearly 70 percent of the women were able to predict which men had papa potential. This was astonishing, considering they had never met the guys. Even more mysterious was how they did it. The giveaway may have something kind, soft, gentler, happier, and, well, paternal in their faces.

Women could also tell which men had the highest baseline testosterone levels. Strong jaw lines (a broad bottom half of the face) and thick facial hair were two reliable giveaways. These men were considered desirable for short-term relationships, whereas the men who appeared the most baby loving were the most desirable for long-term relationships. That said, in the study, there were a few rugged high-testosterone types who were accurately judged as liking children. What we learn here is to trust our instincts. If our first impression of a man is that he's the daddy type, he probably is regardless of stereotype.

Gene police! You! Out of the pool, now!
—Charles Stross, author

IS THERE A DADDY GENE?

The story of the "daddy gene"—or the "monogamy gene"—appeals to those of us who wonder what our partners will be like as fathers. Why is it that some men's hormones and behavior change for the better when they become expectant dads and other men's do not? What makes one guy grow a prolactin paunch and another a beer belly? How can one fellow be a devoted family man and his brother, who grew up in the same household under the same conditions, a deadbeat dad? Is it in the genes?

The quest for answers begins at, of all places, the Yerkes National Primate Research Center at Emory University in Atlanta. There, a biologist, Larry Young, and a postdoctoral fellow, Miranda Lim, have studied prairie voles, a rodent that mates for life. A male prairie vole spends most of his adult life grooming his mate, having sex with her, and building their underground nest. A doting father, he licks his pups, plays with them, and protects them and his mate with his life. He's what researchers call a socially monogamous animal.

Neuroscientists attribute the male prairie vole's fierce fidelity and fatherliness to a feature of his brain anatomy: the vasopressin receptor. Vasopressin is a neurohormone that helps initiate and sustain bonding behavior. It influences the brain wherever there are receptors for it. In the male prairie vole, receptors for vasopressin lushly line the surface of the ventral pallidum, one of the brain's pleasure centers. When a prairie vole male has sex for the first time, his vasopressin levels surge and the hormone binds with the receptors in the ventral pallidum, making him associate that particular partner with pleasure. From then on, he'll defend her and their home with all his tiny might.

Vasopressin also primes the male brain for parenting. When a virgin male rat is given a shot of vasopressin and placed in a cage with pups he never met before, he unhesitatingly assumes the duties of a doting dad. He starts licking and grooming the little ones as if they were his own. (Female voles, like women, are more affected by the chemically related hormone oxytocin.) Both vasopressin and oxytocin are related to bonding and parental behavior.

Young and Lim discovered that male prairie voles are faithful and fatherly because they have a certain variant of the vasopressin receptor gene (AVPR1A). They have what's known as the "long variant." Only voles with the long variant have receptors in their pleasure palace, the ventral palladium.

Size matters here. The short variant of the vasopressin receptor gene is common among the prairie vole's promiscuous cousin the montane vole. These Casanovas don't have vasopressin receptors in their brain's pleasure region. While sex is as enjoyable for them as it is for prairie voles, thanks to the feel-good dopamine that floods their brains, they don't associate the pleasure with a particular partner. You wouldn't expect them to become more attached to their lovers and children than James Bond would.

To test how important the long variant is to fidelity, Young and Lim decided to tamper with prairie voles by blocking their vasopressin receptors. Sure enough, the rodents, formerly faithful and fatherly, became distant and noncommittal after sex. They had one-night stands. Taking the experiment one step further, the researchers transferred the long-variant genes to the brains of the promiscuous montane vole. Again, the transformation was dramatic. Once players, the montane voles become faithful homebodies. Cynics compared them to Stepford husbands. One alteration of the vasopressin gene variant made all the difference between papas and playboys.

Here's where it gets really interesting for us humans. It turns out that men too have different variants of the AVPR1A gene. Some guys are more like prairie voles, and others are more like montane voles. Is there any connection between a man's gene variant and his family life? This is the question that intrigued Hasse Walum, a postgraduate at Sweden's Karolinska Institute. In Walum's study 550 male twins and their partners answered questions, some of them intrusive, about their relationships: "How often do you kiss your mate?" "Have you ever regretted getting married/moving in?" "Have you discussed a divorce or separation with a close friend?" "Rate your degree of happiness in your relationship on a scale of one to seven." Then he tested their vasopressin receptor genes.

What Walum discovered was stunning: focusing on one particular variant thought to be influential in human monogamy and paternal behavior (allele 334), he found that the more copies of it a man had, the weaker his bond with his partner. Men who lacked the gene variant were generally happiest in their relationships—only 15 percent of them had a crisis. Men with one copy of the variant were slightly more likely to report marital problems. And men with two copies were, on average, twice as likely to have had a relationship crisis in the past year than men who didn't have the variant—meaning that 34 percent of them, or one in three, were headed for a breakup.

Their wives and girlfriends agreed. If a man carried one or two copies of the allele 334 variant, his lady was likely to be less satisfied with him than if he had no copies of it. Walum found that a man with two copies is nearly twice as likely not to marry his partner (often the mother of his child or children) as men who had no copies of the variant. Going by these results, there seems to be something suspiciously different about the brains of men who struggle in their role as partners and fathers. They may, due to the type or placement of vasopressin receptors they inherited, have more difficulty being committed family men.

Many of us are now scheming to get a vasopressin receptor gene test for our mates. Of the more than five hundred women who responded to my online poll on this topic, nearly 65 percent said they would test their man if given the option. (You can order a test for less than one hundred dollars online.) In theory, we all want a man without the variant. Here, less is more.

But there's a caveat. Even if there's a correlation between this particular gene and a man's behavior, it doesn't account for all men. Just as the "god gene" and "gay gene" are met with skepticism in the scientific community, so is the "monogamy gene" and the "daddy gene." Even within Walum's study, there were men who had two copies of the variant and were loving husbands and fathers, and there were men without the variant whose relationships were verging on a collapse. The statistics apply to populations, not individuals, who are also influenced by parental role models, partner choice, the opportunity to cheat, past loves, age, life satisfaction, religion, hormone levels, and so on. You can modify the brain and hormones through experience. Put a high-testosterone meathead in the company of Buddhist monks or little girls in pink tutus, and his levels plummet. A man with two copies may become a number one husband and father under the right circumstances. And a man without the variant can also deviate. The gene is predictive, but tendency isn't destiny.

In the meantime, it's fun to speculate: Is it possible that someday we'll have a gene therapy to deliver the right vasopressin genes to the reward areas of the brain, making men more likely to be doting partners? After all, we can do it in voles. Oh, but even then, we can't predict who'll be a faithful father—and who'll be a cheating rat.

> It is much easier to become a father than to be one.
>
> —Kent Nerburn, author

IS PREGNANCY NATURALLY A TURN-OFF?

Between the malodors and the malaise, the fatigue and the fickle stomach, you might not expect to have a vigorous sex life. The deed is done, after all. So what's the point?

Flipping through the illustrated pages of *Pregnant Sex* by Rachel Foux, you'd understand. These ladies are hot mamas. One rides her pardner cowgirl style, her breasts and belly bulging like an udder. Another lady, her nipples like rockets and her distended belly a constellation of blood vessels, grazes red nails over her engorged genitals as she performs oral sex on one partner and takes another from behind.

"With your round tummy and full breasts," Foux gushes, "there's no better time to explore fresh ways of turning each other on, especially if they somehow 'complement' your exciting 'new' condition. Never let those disapproving voices win, the ones that tell you that you couldn't possibly do outrageous things in pregnancy."

Is Foux crazy?

Her book is not the only indication that I could, and perhaps should, be having the best sex of my life. In the back of Dr. Miriam Stoppard's *Healthy Pregnancy* is a chapter on sensuality. If the *Pregnant Sex* gals are out of control, Stoppard explains why: "As hormone levels rise, so does sexual excitement and, incidentally, the amount of estrogen produced in a *single day* is equivalent to that produced by a non-pregnant woman in three years." What follows reads like scientific soft porn: "[We experience an] increase in blood flow to the pelvic area, slightly stretched and swollen labia and vagina, hypersensitive nerve endings, rapid arousal during intercourse, enlarged breasts, sensitive and tingling nipples, profuse vaginal secretions, increased readying for penetration, quick, intense and prolonged orgasm." The list goes on.

So what's wrong with me? At twenty-something weeks, I don't see myself making any advances. Thank goodness I'm not alone. According to a Canadian study of pregnant women of various ages and backgrounds, nearly 75 percent report a decrease in sexual activity, and only 6 percent report an increase. Breaking the numbers down further, nearly 50 percent of women report a plummet in the first trimester, 75 percent in the second trimester, and 76 percent in the third trimester. And what pregnant woman can't identify with at least some of the reasons for the lack of sexual desire: a decrease in vaginal lubrication, sore breasts, cramps, infections, accidental urination, and bleeding? Some women worry that intercourse leads to miscarriage or early labor, or otherwise may harm the fetus. It adamantly does not in a pregnancy that is not high risk.

Many pregnant women (25 to 50 percent) report feeling less physically attractive and think their new body shape turns off their partners. Interestingly, women who are carrying sons have *less sex* than those expecting daughters. No one knows why except to speculate that they may gain more weight and feel worse about their appearance.

The perception that we're not physically attractive is where we usually go wrong. Sure, individual men may be put off by the physical changes or the idea of putting their penis in the proximity of the fetus, but it's less common than we might imagine. In my third trimester, a man who looked ten years my junior hit on me in the cereal aisle at the grocery store. I was suspicious, then astonished. In the Canadian study, 60 percent of men maintained the same level of perceived sexual desire for their pregnant partners, and 27 percent of men actually expressed increased desire. Wow. For most men, the idea of pregnancy sex is a turn-on.

There's an important point to make here. From an evolutionary perspective, sex during pregnancy is a hallmark of our species. While many other animals mate only when females are fertile during estrus, we get it on throughout the menstrual cycle, throughout pregnancy, and well into old age. One explanation is that sex strengthens relationships, or pair-bonding as evolutionary psychologists call it. Pair-bonding is good for the species. As anthropologist Jared Diamond points out, a helpless baby, especially one born in our harsh ancestral past, would hugely benefit from the resources of two parents. If couples didn't literally pair-bond during pregnancy, the male might abandon the expectant mom for more receptive cave gals. Simply put: fathers can be useful, and sex keeps them around.

In both men and women, the quality of the relationship is related to sexual satisfaction during pregnancy. At some point in the nine months, many women have a high sex drive and the strongest orgasms of their lives. We feel emotionally closer to our partners at this time than ever before. Marital satisfaction peaks in the last trimester of a first pregnancy.

But when it comes to naked pair-bonding, the psychological hurdles can be just as high as, if not higher than, physical ones. This is true for me, and I think it's true for most other pregnant women concerned about their body image. Our bodies may desire sex, but our minds hold us back. To turn us around, I think novelist Isabel Allende put it best. "The G-spot is in the ears," she wrote, "and anyone who goofs around looking any farther down is wasting his time and ours."

The other evening my husband said, "I just think you're so beautiful. You look so sexy."

"That's exactly what the baby's father should say," I replied.

"Well, I'll say it, too," he joked, and gave me a feral smile. I laughed. Loosening up made all the difference.

It's for this reason that I root for the improbable babes in *Pregnant Sex*. I admire their spirit of self-acceptance and experimentation. The baby isn't the only thing below the belt.

Moreover, sex is the perfect excuse to do Kegel exercises. (One study found that among women who rhythmically squeezed and relaxed their pelvic floor muscles several dozen times daily over the course of the pregnancy, only 16 percent experienced "muscular laxity" and incontinence after giving birth.) If intercourse is too painful or orgasm too intense, fall back on foreplay, mutual masturbation, oral sex, safer positions, toys, kissing, and cuddling.

The instinct could be there; perhaps all we need is a prod. Cowgirls, ride on!

Could Sex Prevent Miscarriage?

Here's an unusual incentive to have sex during pregnancy: it may prevent preeclampsia, a high blood pressure condition that can result in miscarriage between the middle of the second trimester and the beginning of the third. Preeclampsia warning signs are high blood pressure, severe abdominal pains, swollen limbs, and high protein levels in urine. One in ten women get it.

And here's the weird reason that sex might prevent preeclampsia. Every time you have sex with a man, you're exposed to proteins and hormones in his semen. If your immune system is unfamiliar with the concoction, it may attack the placenta, which has the father's foreign proteins in it, resulting in preeclampsia. When the system is familiar with these proteins—which happens if you have frequent sex during pregnancy—it doesn't attack the placenta as it might after exposure to "unknown" semen.

One study found that if a woman becomes pregnant within the first four months of a relationship, her risk of developing preeclampsia is twelve times higher than if she had been with her partner for at least a year. This is evolution's dark side, according to evolutionary psychologists Gordon Gallup Jr. and Jennifer Davis. The body, when exposed to unfamiliar semen, may assume the timing of the pregnancy is not good. But repeated sex before and during pregnancy has what researchers call a "partner-specific protective effect." It's fascinating.

preeclampsia, which is unique to humans, may have contributed to increased monogamy in our species.

Interestingly, a woman's risk of being struck with preeclampsia also increases if she has several lovers while pregnant or if she and her long-term partner used condoms up until a couple months before she got pregnant. The risk is also higher for a woman who has had artificial insemination with donor sperm as opposed to her long-time partner's sperm. Prepping the woman's immune system with repeated doses of sperm before and after conception may reduce the incidence of the disease.

Does pregnancy sex reassure our bodies that the baby's biological father is still in the picture and will therefore be around to help support the baby? It's an interesting idea. If intercourse during pregnancy seems unappealing, a Dutch study found that oral sex is equally effective against preeclampsia. It turns out that the chemicals in a tablespoon of semen can be absorbed just as easily through the soft tissues of the mouth as the vagina. Either way, it's an ounce of prevention.

5

MAMA'S BOYS, GREEDY FETUSES, AND WHY EVERYONE THINKS THE BABY LOOKS LIKE DAD

On Genes and Biases

One Sunday evening not long ago I saw a comedy routine about a married couple. It started with a man and woman making out on stage. Coming up for air, the wife, wearing a pink bow in her hair, said, "I love you so much! Let's have a baby!" Her suspender-wearing husband beamed and nodded emphatically. "Yes, and I'll buy the baby a pony!" A slight shadow fell across the wife's face. "Well, we don't want a spoiled child." The man, still jolly, snapped his suspenders. "But I'd insist!" he said. The wife put on a sulky and glowering look. She yanked the pink bow out of her hair and wound it around her finger. "You'd be too indulgent a father," she murmured and gave him a hard look. The man maintained a hard, bright smile. "You'd be too cruel a mother." This back-and-forth went on as the couple forecast their future: family therapy, divorce court, custodial fights, a heartbroken child. Finally, the couple looked at each other, exhausted.

"Let's divorce now before it's too late," the woman said. The man nodded, relieved.

I confess the only part about parenthood I'm not looking forward to is the conflict. The squabbles and the skirmishes. The tensions between mother and father, parents and children, grandparents and parents, not to mention sibling rivalry if we have more than one. The accusations of favoritism, cruelty, incompetence, indulgence. The battle for power. The juggling of self-interests.

Problem is, the melee is in our DNA. Literally. These conflicts start from the moment of conception. They're being played out in our genes.

Once upon a time, we thought conception was all about collaboration—at least on a genetic level. We pictured a baby getting a duplicate set of genes from her parents (except for the ones sitting on sex chromosomes), and any given gene would behave the same way no matter which parent it came from. Now we know this is true of most genes, but not all. Sometimes it matters from which parent a gene hales. Certain genes express themselves differently when they come from the father than when they come from the mother, and vice versa. This is called genomic imprinting.

Imprinted genes have a tag that says whether they come from Mom or Dad. The imprint silences the gene. Some genes are always imprinted when they come from the mom, so only the father's gene is active. Other genes are always imprinted when they come from the dad, so only the mother's gene is active. Of the twenty-five thousand genes in the human genome, around one hundred are imprinted. That's less than 1 percent of the genome. But these stubborn, uncooperative, unilateral, one-sided genes have a tremendous impact on pregnancy and the lives of our children, as we shall see. (Imprints are epigenetic; these tags don't affect the underlying sequence of DNA.)

What's also fascinating is how genetic conflicts manifest in how we behave, and especially in how we treat one another. Our genes have their own self-interest, and so do we. Parents and grandparents shouldn't play favorites, but they often do, unconsciously. One parent shouldn't dominate the other in the body or mind of their child, but it happens. Fathers should trust mothers, and mothers should trust fathers, but we don't always—and sometimes for good reason. The human condition is fraught with conflict. It's in our code.

What can we do about it? We can predict the whole drama of life— the trials, the tears, the tug-of-war—but we can't divorce ourselves from it. It's who we are.

We all grow up with the weight of history on us. Our ancestors dwell in the attics of our brains as they do in the spiraling chains of knowledge hidden in every cell of our bodies.
—Shirley Abbott, memoirist

WHAT MAKES THE FETUS GREEDY?

At the table, I attack my husband. "It's you," I say in a wavering voice. I have one arm on the small of my back and the other pressed under the slope of my immense belly. "Just look what you're making me do!" I point at the wrappings and remnants of all that I had just consumed: mashed potatoes, pasta, pound cake, and peanut butter cup ice cream. Peter smiles kindly. "Poor you," he says. "Rough day!"

I can't be angry with the baby. Fetuses need to be greedy. It's her job to manipulate my physiology and metabolism to feed herself and grow. Fetuses commandeer nutrients from our blood and bone. Via their placental ambassador, they homestead in our wombs, digging their tiny heels in maternal soil for as long as possible. They'll take as much as they can get from us for as long as they can get it.

Who's to say how much is too much? We moms do. And who's to say more is more? Dads. The dad's genes just take and take.

Here's what happens. The placenta is not only the baby's ambassador, but the father's too. Though the placenta is made up of both the mother's and father's genes, more of the father's are active, and so the father has the upper hand. Many of the maternal genes are imprinted, or silenced, according to Harvard University molecular geneticist David Haig, the founding father of the field known as genomic imprinting.

Paternal genes in the placenta are bent on amping up the feed controls. The sneaky placenta achieves this by releasing a hormone called human placental lactogen, which, by week 20 or so, reduces a mother's sensitivity to insulin and raises her blood glucose level. Lactogen gives us the appetite of lumberjacks. The more sugar in our blood, the more the greedy fetus can feast on. If the sugar levels escalate out of control, we get gestational diabetes and the fetus's metabolism goes awry.

As terrible as this sounds, most of the time the conflict between Mom's and Dad's genes, as Haig describes it, is "no more than a petty squabble over precisely how much glucose and lipid the fetus receives." A peace pact is eventually made. We dodge some of the fetal-paternal demands and cave in to others (which can be considered practice for later). To the extent that we give in and eat more, the placenta gets bigger, the little one gets more nutrients from us, and the pregnancy may possibly drag on longer (see "Is Daddy Delaying Us?" page 167).

Those pushy paternal genes in the placenta evolved long before we became modern humans, but we can imagine the conditions that made them beneficial. They hail from a time when mothers and fathers had very different interests (or at least their genes did). Couples were not monogamous. Food may have been scarce, conditions harsh. Pregnancy was a major investment (it still is). Reproductively, it was in a woman's interest to budget the amount of nutritional resources she could spend on each pregnancy. The body should not be overburdened. This tactic maximizes the number of healthy children a woman could have in her lifetime.

This strategy didn't sit well with the paternal genes. From an evolutionary perspective, it's in the father's best interest for the fetus that is carrying his genes to survive and pilfer the most resources possible from the mother. From the perspective of his selfish genes, why should the father care about the mother's present and future sacrifice? No sweat off his hairy back. If this pregnancy doesn't work out, he has no guarantee that his genes will get another chance, so he needs to feed this baby as much as possible. Times have changed (well, maybe), but these are the conditions under which our ancestors evolved.

Feeding the fetus isn't the only act of paternal genes indulging the fetus at the mother's expense. After the baby's birth, they drive the baby's appetite, metabolism, attention demands, and postnatal growth rate. Fathers, my dear husband included, tend to take in this informa-tion with a certain pride. Some men are particularly pleased to learn their genes urge their infants to suck harder at the breast. They inhibit appetite for supplemental food at the time of weaning. We know this because babies born with absent or nonworking genes from the father (on chromosome 15) have a condition known as Prader-Willi syn-drome. These babies are born slack-mouthed; they can't suckle or swal-low well. As a result, they have developmental delays and poor weight gain early in life.

All this paternal aggression isn't for naught. In the case of the pla-centa, the sugars it works to extract from me strengthen it as well as the baby. The human placenta needs strength because it burrows an incredible one-third of the way into the wall of the uterus, more than in other mammals. The deeper it digs, the more resources it accesses. Without this deep drill, the baby's brain wouldn't get enough. We sac-rifice for a higher purpose.

Which is exactly my excuse when the placenta makes me ravish another chocolate bar. It's for a higher purpose.

Who Controls Baby's Brain?

Your and your partner's voices, encoded in your child's DNA, tussle for control. For imprinted genes, some of which influence brain function, there is a significant bias as to which copy is active: the one inherited from the mother or the one that came from the father.

Right now, in midpregnancy, I may have the upper hand when it comes to my baby's developing brain. In a recent study of mouse embryos at Harvard University, Catherine Dulac and her colleagues found that about 60 percent of the active genes at this stage are maternal in origin. Since many genes in rodents have human counterparts, the researchers suspect that a substantial (yet lesser) number of imprinted genes are also in humans. How marvelous, I think. This would mean my maternal genes are laying the foundation of my daughter's mind.

I'll savor the power while it lasts. By the time a baby reaches childhood, maternal genes have less and less influence over mind matters, according to the same study. Paternal genes that had been passively standing by begin to switch on and take over. Later, in adulthood, the father's genes get the upper hand, and a mom must accept that more of her genes are silenced. About 70 percent of the active genes in the mouse adult cortex and hypothalamus are paternal (excluding the X and Y chromosomes). There's a fascinating theory that autism spectrum conditions strike when paternal genes are overexpressed and raise their voices too high, and schizophrenia strikes when maternal genes do.

Parents influence the brains of daughters differently from sons, which is where the X and Y chromosomes come in. A boy gets his only X chromosome from his mother, so Mom dominates when those genes are active. We arguably have more sway in a son's brain than in a daughter's. Daughters, with their X chromosome from each parent, may have three times the number of silenced genes as males do. Mysteriously, a father's X-linked genes appear to have the upper hand in his daughter's preoptic area of the hypothalamus, the region that governs social life, mating, and maternal behavior. (Could my dad's genes really have more say in who I'd choose as a partner and how good a mother I'll be?) However, the maternal X-linked genes are louder in the daughter's cortex, a brain region that plays a key role in memory, attention, and consciousness. If my daughter someday forgets to call, I'll have our own genes to blame.

Or I'll just have to accept that she actually has a mind of her own.

WHAT GENES DO BABIES INHERIT FROM MOM ONLY?

Late in my second trimester, I join two dozen other pregnant women in the candlelit sanctuary of a prenatal wellness class in downtown Manhattan. We are all in our late twenties to late thirties, a smattering of writers, artists, lawyers, and Wall Street types, and all in various stages of pregnancy. We wear elasticized yoga pants and our hair in ponytails or yoga knots. Our paunches oiled, we sit cross-legged as Willow, our instructor, begins a guided meditation.

"Imagine you're on a long journey," Willow whispers. She resembles a pilgrim: witchy hair, beet-dyed sack dress, fingerless lace gloves, bulging eyes. "You're climbing a mountain," she says. "And on the trail, at the side of the trail, you see your mother. You hug her. She has something for you, and you take it."

From the back of the room I hear a snicker, but Willow perseveres.

"You keep seeing other women at the side of the trail. Next is your mother's mother, your maternal grandmother. You hug her. You keep climbing and encounter her grandmother, your maternal great-grandmother. She you also hug. As you climb the mountain, you'll meet women who are both strange and familiar to you: your maternal great-great-grandmother, and her mother, and her mother, and so on. Each woman bears a gift for the baby. The gift of joy. The gift of courage. The gift of wisdom. The gift of laughter. The gift of grit. You accept them all on behalf of your baby."

Willow finishes the meditation with a long OMMMM and smiles broadly. "This, ladies, is your maternal line. A chain of mothers and daughters going back through all of history."

Willow's meditation makes me think of mitochondrial DNA (mtDNA), which resides on the X chromosome and is passed down from mother to child over the generations. Mitochondrial DNA is special among genes because it's inherited intact and exclusively through female bodies. Unlike most other genes, *only* the mother's genes have a voice here because the father's are absent. So unaltered is mtDNA over the generations that we can each trace our lineage—mother, mother's mother, her mother, and so on—for thousands of years. Women pass these genes to sons and daughters alike, but only a daughter can pass them on to the next generation. Sons are a dead end.

What mtDNA does is as inspiring as what it represents. These ladies are powerhouses. Hundreds of copies of mtDNA are in almost every cell of the human body. They code for mitochondria, tiny organelles that convert energy into a form the cell uses for its growth and function. Mitochondria are densest in parts of the body that need the most energy: the muscles, liver, and, intriguingly, the brain. Because mtDNA is the brain's power supply, any alteration to it affects how the entire mind functions. Research suggests that these maternal genes play a role in extroversion, mood disorders, and longevity.

Early in my pregnancy, when the embryo must've been no more than several hundred cells, long before I was confident it would stick and become viable, I tried to visualize the mtDNA at work. It's as if me, my mother, my mother's mother, her mother, and so on—generations of quirky, addled Ashkenazi women and their foremothers—were cheering the little embryo on, driving her development and ensuring her survival. "The gift of endurance," as Willow would put it. Now that I know the baby's a girl, she'll pass on mitochondrial genes to her own children if she chooses to have them. The voices of our foremothers would echo on for another generation.

Because mtDNA is so hardy and faithful, it's useful when trying to figure out who's related to whom, even after thousands of years, by isolating it from the bone marrow or a hair shaft and looking for tiny mutations that act as identifiers. To identify the bodies of the Romanovs, the Russian imperial family, researchers used mtDNA from a sample provided by Prince Philip, who shares the same maternal line. Researchers confirmed Jesse James's remains were his by following his maternal line to the son of his sister's great-granddaughter.

Go back far enough, and you'll see that our mtDNA ultimately originates with ancestors that are not even human. The mitochondria in all our cells, propelling us through our lives, are so ancient that they were once a separate species of aerobic (oxygen-using) purple bacteria. In a sort of marriage of convenience, they joined forces and coevolved inside other, larger cells called eukaryotes. Mitochondria received protection and nutrients from the larger eukaryotes in exchange for enzymes that produce energy efficiently. Before this happened, eukaryotic cells were inefficient; they relied on fermentation and other sluggish processes. With oxygen-fueled mitochondria as a power source, they became more complex and multicellular. It was like upgrading from an Edsel's engine to a racecar's. Fueled by mitochondria, complex life-forms from fungi and fish to human beings became possible.

I find it sweetly symbolic that mitochondrial DNA originates from symbiosis, a relationship of mutual benefit and dependence. It's a lesson a mom might teach: cooperate and everyone wins. It's one of Mother's Nature's great gifts.

Tracking Our Foremothers

Who were your foremothers who gave you and your children your mtDNA? Where did they come from, and where did they go? Molecular geneticists can tell by identifying markers in the mitochondria that distinguish populations as they migrated across the globe. You can trace your maternal ancestry—your maternal line tree—with a service such as Ancestry.com or 23andMe. In exchange for a swab of saliva drawn from your cheek, you'll learn the name of your family haplotype or family branch.

It works like this. Mitochondrial DNA is passed faithfully from mother to child. But every once in a while—usually about every ten thousand years—a tiny mutation creeps in and is passed on to subsequent generations. By analyzing the patterns of mutations in mtDNA, geneticists can trace when and where one group splintered off from other groups. Your mtDNA will help you trace the migration of your maternal-line ancestors only. The paternal line—your father, your father's father, his father, and so on—is tracked the same way via DNA on the Y chromosome passed from father to son.

My maternal line, I learned, is Haplogroup K—inherited from a band of Middle Eastern wanderers who became a distinct population fifty thousand years ago when they traipsed over to Europe.

An mtDNA test doesn't reveal anything about your father's mother or our mother's father, and so on. Go back ten generations, and we have over one thousand ancestors; go back twenty generations and we have over 1 million ancestors, but we inherited mitochondrial DNA from only one. Going further back in human history before the various haplogroups branched off from one another, all of us have a common female foremother—a so-called mitochondrial Eve—who is estimated to have lived about 140,000 years ago, in East Africa, around current-day Tanzania.

HOW ARE ALL SONS MAMA'S BOYS?

While daughters carry on their mother's mitochondrial DNA, boys have their own special bond with Mom. An overlooked biological oddity is that sons inherit more of their mom's genes than their dad's. If a woman has a son, he'll be (slightly) more genetically related to her than would her daughters. He inherits a larger volume of her genes proportionally.

Here's how it works. Drill deep into any of the fetus's cells, and you encounter a tight little knot of genetic material. These are chromosomes. Take them out, pick them apart like a tangle of necklaces, and lay out all the little segments. That's what technicians do when they unwind and map out fetal cells drawn in an amniocentesis (a test that detects or rules out certain genetic disorders). In a normal baby you'd find forty-six chromosomes, twenty-three pairs.

Chromosomes 1 through 22 are the same for males and females (called autosomal chromosomes), and they're always the same size from each parent. Then there's pair 23, the sex chromosomes. A girl has two X chromosomes—one from Mom and one from Dad. A son has an X chromosome from Mom and a Y chromosome from Dad. The Y chromosome doesn't look like any of the other chromosomes. It's puny. Next to the X chromosome it looks like a pygmy paired with an Amazonian queen.

Size matters. Her Highness, the X chromosome, holds several times more genetic information than the Y, and it's physically much larger. There are over fifteen hundred genes on the X, representing about 7 percent of the 20,000–25,000 genes in the human genome. (Many of the X genes are silenced or imprinted, so the chromosome is impressive but not quite as domineering as it sounds.) In contrast, there are fewer than one hundred genes on the Y chromosome. The difference is about 5 percent of a man's total number of genes. Looked at proportionally, in this crude quantitative way, a son gets significantly more genes from his mom than from his dad. And more of those maternal genes are active in him than they are in his sister.

A daughter gets a fairly even distribution of her parent's sex chromosomes. She inherits two X chromosomes, one from her mother and one from her father, and a more or less equal mélange of maternal and paternal genes. In any given cell, one X chromosome is active and one is usually inactive; the decision is mostly random (except for imprinted genes). The genes act like well-mannered dance partners, politely stepping back when their counterparts take the floor. Mom's X genes shut down when Dad's X genes are on; Dad's shut down when Mom's take over.

Not so with sons. The first twenty-two pairs of chromosomes line up and do their courtly exchange. But if the X and Y tried to tango, it's a no-go because hundreds of genes on Mom's X don't have dance partners. Lacking a counterpart, a male's sole X chromosome, inherited from his mom, takes a leading role wherever it's needed—and that's in nearly every cell of his body.

Just about everywhere you look, the X plays a role. One out of every five brain disabilities is related to X-linked mutations, suggesting that X-linked genes affect intelligence and brain function. X-linked genes also strongly influence fertility. Because males have only one X chromosome, they have no safety net if something is amiss. This is why men are more likely to get X-linked conditions, from hemophilia to color-blindness and forms of muscular dystrophy. A mother with a defective X-linked gene has a 50 percent chance of passing it on to each of her children, but only her sons will get the disease while the daughters will be carriers (unless their father has the disease too).

In a daughter, one X comes from Mom. The other, inherited from Dad, is her paternal grandmother's. This means half of a daughter's X chromosomes are her paternal grandmother's. It pleases my husband immensely that our baby girl will inherit his mother's X chromosome through him. This is how Mama's boys pass on their legacy.

> Men are what their mothers made them.
>
> —Ralph Waldo Emerson

DO MEN PREFER BABIES WHO RESEMBLE THEM?

My husband has an idea. "Let's get a 3D ultrasound," he says. "I love information."

I remind him that the two-dimensional ultrasound, the one pre-scribed by our doctor for all but high-risk cases, is sufficient for our purposes—to detect ten fingers and ten toes, a slinky of spinal cord, the absence of cleft lip, and a heart that beats. There is no apparent medical advantage in seeing our fetus in three dimensions.

"Except in 3D we'd get to see her face more clearly," I add noncha-lantly. I think I know what's going on here.

"Oh yeah, that'd be nice," Peter says casually.

My guess is that he is aching to know if the fetus looks like him. Rationally, he knows she's his baby. But this is something primal. I think he's looking for paternal certainty in our unborn baby's features.

Do men really care that much?

This question has piqued the interest of several research psycholo-gists over the last couple of decades. Much of their work has depended on digital tools that can create composites drawn from photos of sev-eral faces. In doing this, they create images of kids that look as if they could be sons or daughters of the research subjects, with varying degrees of resemblance.

One of the most widely known researchers in this field is psycholo-gist Steven Platek. Several years ago, Platek and his colleagues recruited a bunch of men and women, took photos of them, and digitally blended their facial features with those of babies and children of various ages. To assess how much his subjects might unconsciously favor a self-resembling child, Platek showed them several sets of ten photos, one of them digitally manipulated to look like the subject. "Which one of these children would you be most likely to adopt?" he asked. "Which one of these children would you least resent paying child support for?" "Which one of these children would you spend the most time with?" "If one of these children damaged something valuable of yours, which one would you punish the least?" Other questions followed.

The results were intriguing. In this experiment and follow-up stud-ies by Platek's group and other teams, the male volunteers significantly favored pictures of kids that were morphed to resemble themselves. These were kids they wanted to spend time with or support financially. This bias was generally found in men, but not women. It applied to boys and girls of all ages, from babies and toddlers to elementary schoolers.

A male preference for look-alikes is quantitative too. The more a child looked like a man, the more the man liked the child. As the percentage of a guy's features increased from 12.5 to 50 percent in a child's face, the more he was willing to adopt or spend time and money on the little one, and the less likely he was to want to punish him or her. In one study, Platek determined that a child's face had to have a minimum of 25 percent of a man's features for that man to favor him or her over any other. That's the equivalent proportion of genes he'd share with nieces, nephews, half-siblings, and grandchildren.

This makes sense. In an ancestral environment, men didn't necessarily know what their own faces looked like. Like Narcissus, they could gaze into pools to see their reflections, but they had no mirrors. Yet a man knew the faces of his brothers and sisters (who share 50 percent of his genes) and their children (who share 25 percent), and from this he could determine if the faces of his mate's children looked anything like his kin.

Fascinatingly, men's favoritism for self-resembling babies and children is completely subconscious. In one experiment, researchers interrogated the subjects about why they chose certain kids over others. The men would shrug their shoulders. They had no idea. It was just a first impression, and it wasn't grounded in any real reason or understanding. A few said they were just "going with a gut feeling." They didn't detect any resemblance between themselves and the kids they favored, nor did they realize the faces had been blended. When the researchers revealed what the experiment was about and showed the volunteers the digital technology, the subjects were shocked. Only when they were shown their own original picture next to the self-resembling child's face could they consciously see the resemblance.

Men's bias is so subconscious that Platek and his colleagues began to suspect that there's something unique about how the male brain processes children's faces. Following this hunch, they recruited men and women for several experiments involving a brain scan known as fMRI, functional magnetic resonance imaging, which tracks blood flow throughout the brain.

What they found is telling: male brains were much more active than female brains when responding to self-resembling children's faces than when looking at children who didn't look like them. While a woman *knows* a baby is hers, men must be more discerning, and that takes more mental processing. Men's anterior left prefrontal lobe and anterior cingulate gyrus are particularly active when they view pictures of children who resemble them. These regions are normally associated with the inhibition of negative responses. Could it be that men are unconsciously assessing their chances of being biologically related? They may be generally skeptical about children's relatedness to them, but that reflex is suppressed when they see children who resemble them. Women, meanwhile, may be using their brainpower to assess personality cues in children's faces. A woman's investment in a child may be linked to a similarity in personality much more than facial resemblance.

What this experiment shows is that men evolved to be discriminating when deciding whether to invest time and resources in a child. Recently, in a polygamous population in rural Senegal, French researchers found that fathers invested more in children who looked (and, incidentally, smelled) like them. Daddy-resembling kids were taller and better nourished because they benefited from better fathering and provisioning. Fathers favor look-alikes.

I share all this information with Peter, and he laughs.

"I already know she's mine. There's a better test of relatedness than what she looks like," he smiles.

"What?"

"What she does when you eat mango."

It's true. Mangos, one of my husband's favorite foods, appear to be one of our daughter's faves too. She goes wild in the womb when I eat them.

Okay, I think she'd like sweet mango regardless of her paternal origin. But why would I tell my husband that?

> Self-love is the source of all our other loves.
> —Pierre Corneille, French dramatist

WILL THE BABY REALLY LOOK MORE LIKE DAD?

Several decades ago, two of the world's preeminent evolutionary psychologists, Martin Daly and Margo Wilson, a husband-and-wife team, decided to find the truth about whether people think newborns resemble their fathers. Daly and Wilson knew that within minutes after a birth, families often talk about whom the child takes after. To capture these first conversations without interfering or influencing them, the couple relied on video cameras that had been planted in the delivery rooms at the University of Colorado Medical Center. Over one hundred after-birth videotapes were made and each was analyzed and coded according to how many times a speaker remarked on maternal resemblance, paternal resemblance, or a resemblance to someone else in the mothers' and fathers' families.

A typical conversation would go like this:

Mother: It looks like you.

Father: (no evident response)

Mother (a little while later): He looks just like you!

Father: (nod)

Mother (to hospital staff): He's cute. Looks just like Bill.

Father (embarrassed?): Don't say that.

Mother: He does.

Three times more often the father was said to resemble the newborn than the mother. Three times as likely! Sometimes the mother would comment on her baby having her husband's eyes or head of hair. Sometimes the grandmother would be the one to make the comment.

There were mothers who'd say to their partners, several times over, as if to make sure it sunk in and stuck, "The baby looks like you."

So that settles it, you might think. Of course people say babies look like their dads—because they do.

Really?

Only recently has the baby-looks-like-daddy belief been put to the test. Psychologists at the University of California at San Diego recruited volunteer judges to match photos of boys and girls—infants, ten-year-olds, and twenty-year-olds—with their biological parents. Each child was paired with three possible parents of each gender, including the real mother and father. These were actual pictures of parents and their kids, so you'd think matching them would be easy. But it wasn't. Surprisingly, the judges were not able to match the older children with either of their parents any better than by chance. For kids aged one and younger, however, they were slightly more accurate when identifying the right father.

But no one has been able to replicate the results of that study.

Psychologists at the University of Liège in Belgium attempted the exact same experiment with a different and larger set of babies and adults. This time, the judges completely bombed: they had difficulty identifying either biological parent, no matter the child's age. Similar studies in France and at the University of Georgia found that newborn boys and girls actually appear to resemble their mothers slightly more than fathers, even though the moms often insisted that their babies looked more like their dads.

The assumption that kids, especially newborns (or even fetuses in ultrasound scans), more often resemble their dads is so unsupported that most evolutionary psychologists now conclude it can't be true. Sure, biological similarities can be teased out. A given baby may look more like her father than her mother or vice versa. But the truth is that babies look more like one another than they do their moms and dads. The similarity to either or both parents is not that much stronger than it is to random adults of the same race.

But here's the bizarre part: we all actually believe it when we tell a new dad that the baby looks like him. The ruse is subconscious. According to evolutionary psychologist Paola Bressan, mothers are unaware of their deception when cooing about how much the baby looks like her partner and fathers are unaware of being deceived. This is true across all cultures. Even perfect strangers with no ulterior motive have an unconscious paternal bias. If you tell them that a man is a baby's father, as studies have found, they'll see a much stronger resemblance than they would otherwise. (Good news for parents who adopt.)

The best part about all this is that fathers too are more likely to claim paternal, not maternal, resemblance—mostly because they want it to be true. While men seek assurance of their paternity, many also appear not to mind ambiguity. Only 50 percent of men are in favor of routine paternity testing. That's lower than we might expect.

From an evolutionary point of view, says Bressan, the universal tendency for men to see themselves in their infant's features is a lifesaver for infants because it's a face-saver for cuckolded men. Male chimps are known to abuse or kill babies that aren't theirs; humans may have similar instincts. If paternity really were obvious, the father—and everyone else—would not only be certain when the child is his, but also when it is not his. Given that approximately 3–5 percent of babies are not genetically related to their presumed dads, many would be mistreated, abandoned, or worse by furious cuckolds made aware of their status. An interesting thought: Are we all unconsciously protecting infants by claiming that they look like their dads, knowing intuitively that men prefer babies who resemble them?

Yes, it's better for the survival of our species if babies are blank screens onto which we project a paternal bias, even if the bias is flawed. Paternity, for a good reason, is in the eye of the beholder.

Mama's Baby . . . Daddy's, Maybe
—Anonymous

DO GRANDPARENTS (UNCONSCIOUSLY)
PLAY FAVORITES?

Not long ago, a group of evolutionary anthropologists invited men and women to participate in a study. The subjects were given a list of questions about how they feel about members of their family. The researchers told the volunteers to be objective and that the results would be anonymous. They asked them to think of their four grandparents: maternal grandmother, maternal grandfather, paternal grandmother, and paternal grandfather.

"Which one are [or were] you closest to?" the researchers wanted to know. They asked their subjects to think hard about who was the most influential in their lives, including how many gifts and how much emotional support they received.

For me the answer is easy. My maternal grandfather comes first. Grandpa E. had a weakness for desserts and lived in a house full of Chinese antiques and brass-framed photos of me and my brother. When I think back, I can recall the smell of prime rib and coffee cake—so much of our time was spent eating—and the hushed tenor of his voice, in a faraway room, advising my mom on child rearing. This was a man who mailed me weekly letters in his crabbed scrawl, offering life lessons: I shouldn't marry too soon or get pregnant too young, and I should work hard.

I'm not alone in my maternal-side bias. Male or female, participants in many studies say they identify with, are emotionally closest to, receive more resources from, and more often see their mother's parents than their father's. Most rank their grandparents in the following order: maternal grandmother, maternal grandfather, paternal grandmother, and paternal grandfather. All else being equal—resources, geographical proximity, affection for their children, and so on—the mother's parents are often emotionally closer and contribute more to their grandchildren than the father's parents do. Even when they live far away, maternal grandparents, especially maternal grandmothers, are more inclined to maintain frequent contact with their grandchildren. They go the extra mile. Obviously this isn't true of all families. We're talking trends here, not individuals.

This pattern of maternal grandparent favoritism, with Mom's mom ranked first, has been found in dozens of studies worldwide. This is also true of other relatives; maternal aunts, especially a mother's younger siblings, outrank paternal aunts and, especially, paternal uncles. My maternal grandmother passed away before I was born, but my maternal grandfather played the grand role, and I'm also close to my maternal uncle, my mother's only sibling.

I was amazed to learn that a matrilineal bias usually happens unconsciously on the part of both grandparent and grandchild. Some interpret the caregiving bias as a natural extension of family ties and the way women relate—the bonds between mother-daughter and aunt-niece naturally extend to the younger woman's children. Since a mother needs more direct help in raising children, it follows that the maternal line may be favored. It may also be in the grandparent's best interest to favor grandchildren born to the child most likely to take care of them in old age, and that child is more likely to be their daughter than their son.

While all these theories play a role in maternal-line favoritism, evolutionary psychologists argue that there's a darker motivation: maternal certainty versus paternal uncertainty. *Mommy's baby, Daddy's maybe*—we always know who the mother is, but who knows about the father?

Fact is: if you're a grandparent, especially a grandmother, you know your daughter's kids are biologically related to you. There's certainty that those children have about 25 percent of your genes. On a subconscious level, you may even favor your daughter's children based on a biological signal of genetic relatedness such as smell. Do maternal grandchildren have a different "odor print" from paternal grandchildren? I'd be excited to see this research; there may be a bias here too.

Take a moment to consider the plight of paternal grandparents. For all they know, their daughter-in-law could have taken on a lover and your grandkids' biological dad is the body-builder bachelor next door. Kinship is doubly uncertain for the paternal grandfather, who may be twice removed: if he's cuckolded by his wife and his son by his daughter-in-law. This may be why Dad's dad often gets the lowest ranking by his grandchildren.

So the star here really is the maternal grandma. Why? If she's alive, grandkids survive. From Europe and the Americas, to the highlands of North India and the lowlands of Asia, to the outback of Australia and the backwaters of Gambia, anthropologists have found a "maternal grandmother effect." They've seen her digging for tubers with her daughter's newborn strapped to her back, working her long-standing social connections, brokering sibling disputes, wiping away tears, matchmaking, and doling out advice. Her grandchildren are taller and better nourished than children whose maternal grandmothers are not present. Women who have their mothers around to help out also tend to have healthier, heavier babies at birth. They may have larger families because they can wean each child sooner, thanks to food that the grandmother provides that they wouldn't have been able to obtain otherwise. Those babies are more likely to make it to adulthood, find mates, and pass on their grandma's genes to future generations. (Love isn't the only motivation for these grannies; in ancestral settings, whether they were buried alive or abandoned depended on their usefulness.)

The father's mother can be helpful too, but not generally as much as the mother's mother. The involvement of a man's mother is linked with her daughter-in-law's getting pregnant soon after marriage and having more children in rapid succession. This may sound like a good thing, but it often leads to a higher rate of stillbirths and newborn mortality, presumably due to the mother-in-law's nagging and harping. Only in rural, patriarchal societies are paternal grandparents more involved than maternal ones, and that's only because women move in with their husband's parents and only sons inherit land and resources.

When grandparents themselves are asked to rank their closeness to their sons' or daughters' children, they usually don't express any bias. It doesn't seem right or fair to pick favorites. When I mention the maternal bias to my own mother, she agrees that there's something special about her daughter having a child. For years she had been coaxing me to give her "her grandchild." But favoritism for my babies over my brother's? No, never. She loves all her grandkids.

———

Reflecting on Grandma's contributions to the human race, evolutionary psychologists have decided she deserves credit for giving humankind two gifts: intelligence and longevity. The gift of intelligence comes from the time she spends helping raise grandchildren as their big, slow-maturing brains develop. Without that help, our species would be under pressure to mature faster and miss out on all those years of brain building. Longevity, the other grandmotherly gift, comes from the evolutionary pressure on women to live longer to help out their families. Women with "longevity genes" have had more surviving kin than their short-lived peers, so longevity genes have propagated among us. Only whales and women live decades after they can no longer reproduce.

What about Gramps? According to many evolutionary theories, men go along with women for most of the ride into old age. Although grandfather presence hasn't been linked directly to grandchildren survival rates as grandmother presence is, that doesn't mean Gramps is useless baggage. The Methuselah-like patriarch helps his kin by protecting them and promoting their long-term interests. If his children and grandchildren are able to thrive and have lots of kids, then his long-lived genes persist. And if that patriarch is a virile old chap, he may have contributed to human longevity in other ways too: the willingness of young women to mate with powerful old men perpetuates longevity genes as well.

Long live grandparents. They made us who we are. And we may become them someday.

6

FRAZZLED FETUSES, SNOOPING GENIUSES, AND WHY CHOCOLATE LOVERS HAVE SWEETER BABIES

Prenatal Predictors

At week 32, the doctor tells us that our baby is breech. While 85 percent of her peers are head down in their mothers' pelvises, all teed up to be born, our baby is dancing and dreaming and dawdling. Her big round head bobs around my belly button. One of her feet is crunched against her belly, and the other is tapping my cervix. My obstetrician informs me that a breech birth is potentially dangerous because the baby, navigating the birth canal butt- or feet-first, may get trapped or the umbilical cord compressed. Somewhat controversially, women in the United States are routinely scheduled to have a Caesarean section if their babies are still breech near their due dates.

I try everything to get my baby to turn in the head-down vertex presentation while there's still time. I lie on the floor with my feet on the sofa so that my hips are at least a foot above my shoulders. I burn a stick of mugwort near the pinky toe of each foot for twenty minutes twice daily—an ancient Chinese technique to turn a fetus. I do somersaults in a glittering lake in northern Massachusetts. I place a bag of frozen peas over her head and a warm buckwheat sack down below, following the suggestion that fetuses, like migrating butterflies, are attracted to warmth. But the baby won't budge.

My husband and I see this as a sign of an emergent personality.

"Stubborn, just like her dad," I say.

"She's just going in her own direction," Peter explains. "In her own sweet time."

There's no real reason that some babies are head up when they should be head down. Personality almost certainly has nothing to do with it. But here we are, reading the runes of the ultrasound.

Like most obsessive expectant parents, we have been extrapolating the little information we have to predict our baby's character and temperament. Every scan, every kick, every flutter is a sign. Her size, her sleep cycles, her season of birth—everything is taken into consideration. Will she be an extrovert, a worrywart, a bully, or a klutz? Sweet-tempered, feisty, or morose? Will she be a night owl or a morning bird? A neurotic? A daredevil?

Who is she, and what is she thinking?

By now a lot has happened inside that big skull we see in the ultrasound. Months ago, back in the first trimester, construction of her brain began with a foundation of cell layers called the exoderm and the mesoderm. Together they formed a neural plate, which became a neural tube, which became her central nervous system. Neurons are forming in her brain at a rate of 250,000 a minute and migrating like pioneers. Once a homestead was established, they laid down long, thick fiber cables of axons—information superhighways. The axons are linking up. They form networks and patterns of flow.

By week 6, a fetus's first brain waves are detectable. The neural tube breaks out into a forebrain, a midbrain, and a hindbrain. The hindbrain is the physical plant, home of the medulla (mediates breathing, swallowing, heart rate) and the pons (eye and body movement). The hindbrain and midbrain together give rise to the cerebellum, which coordinates movement and balance.

By week 9 a fetus can move, and her heart beats. She can swallow, sigh, stretch, and suck her thumb.

By week 10 she has a preferred hand, left or right. She starts to develop fingerprints.

By week 12 she can open her mouth in response to stimuli.

By week 14 she responds to touch.

By week 15 twins start to massage and mimic each other in the womb (we're social well before birth).

In the second trimester, the grandest architecture by far, the forebrain, begins to develop. This is home to those hallowed structures, the diencephalon and the telencephalon. The diencephalon develops into the thalamus (relays sensation and motor) and the hypothalamus (controls emotions and sensory perceptions and the release of hormones). From the telencephalon, the hippocampus is formed (long-term memory and spatial navigation). The hippocampus will be nearly complete by the time the baby is born. The forebrain is also home to the basal ganglia (controls movement, sensory information) and the amygdala (attaches emotional significance to the signals that it sends to all the other regions).

By week 23 a fetus can hear, taste, and see (somewhat).

By week 26 she can recognize her mother's voice.

In the third trimester, the prefrontal cortex is ready for ribbon cutting. It's the esteemed realm of learning, language, and abstract thought. It is here where 70 percent of the nerve cells reside. This area will continue to grow long after the baby is born.

By week 27 a fetus begins to dream.

By week 32 most fetuses are in the head-down, or vertex, position for birth.

And at week 38-and-a-half, our fetus spins out of breech.

This happens after I play Mozart sonatas for two hours through a set of small speakers squeezed between my thighs. My husband and I had been laughing and enjoying the absurdity, and then I felt a watery shift within. The next day, our doctor confirmed the miracle: the baby's head was in my pelvis. And so was her hand, reaching down as if she had been trying to touch the music.

"She loves classical music! What refined tastes!" Peter and I coo.

Oh, here we are, projecting again. Determining her disposition, shaping her sensibilities. Prenatally predicting.

The truth will soon emerge. Head first now. And as the time nears, I keep wondering: *Who are you? Who will you be?*

> The best predictor of future behavior is
> past behavior.
>
> —Dr. Phil

WHAT DO FETUSES LEARN BY EAVESDROPPING?

Our babes can snoop on us by third trimester. I sense mine does. The nosy little busybody, she's listening to me chitchat and argue and grouse. Noises are distorted in her watery bubble, but she hears me clearly enough. My voice is louder than anything else in her world. My sound surrounds her. She swims in it. It'll probably never influence her more than it does right now.

At this moment we're at 30 weeks, the age at which a fetus is nearly able to process and prioritize sounds. When I speak, her ears should perk up. If I were to play a recording of myself reading *Bambi* through a loudspeaker suspended above my belly—as mothers in a study at the University of Toronto did to their fetuses between weeks 33 and 41— her heart rate should speed up by about seven or eight beats per minute. This would mean she is aroused and paying attention. But if an unknown person were to read the same passage, my unborn baby's heart rate would remain unchanged.

Fetuses can learn which background noises are normal and tune them out. This is evidenced by a heart rate that neither speeds up nor slows down. Infants whose mothers lived near Osaka International Airport in Japan before and during the third trimester of pregnancy don't lose any sleep when a recording of aircraft engines roars over their cribs, but they're terrified by the tink-tink-tink of piano keys. My city fetus probably wouldn't startle from the sound of sirens and screeching brakes. She's hardened even before she's fully baked.

How do we know that a fetus (from about 30 weeks) not only listens but remembers what she hears after she's born? Here's a simple experiment. Read a book out loud—*The Cat in the Hat*, for instance—to your abdomen twice daily in the third trimester, as the expectant moms did in an experiment led by psychologist Anthony DeCasper at the University of North Carolina. Once the baby is born, read a few books to him, *The Cat in the Hat* among them, and see if he appears to favor Dr. Seuss. DeCasper found that newborns did if they had been prenatally exposed to the book, as evidenced by their sucking on a nipple connected to an audio device. The machine would play Mom's voice reading the Seuss book only if they sucked at a certain rate. Otherwise, they would hear their moms read a different book. Fifteen out of sixteen babies sucked at whatever speed was required for the machine to play *The Cat in the Hat*. The story didn't mean anything to them yet, but they remembered how it sounded when they heard it in utero. They remember who read it to them. And what's familiar is preferred, at least at this impressionable age.

A mom's voice plays a special role in her baby's acquisition of language. In one Canadian study, newborns—literally, babies who were only hours old—wore electrodes on their soft furred scalps so that a researcher could track their brain activity when hearing voices. When an unknown woman made a sound with a short "a" vowel sound, as in *cat* and *hat,* the newborns' right temporal lobes would light up, associated with recognition that a sound is a human voice. This is somewhat interesting; it proves these regions of the brain are online at birth. But when a baby's own mother made the same sounds, the newborn's left posterior temporal lobe became activated. This is striking—because that's the part of the brain associated with speech and language comprehension, and only Mom's familiar tone seems to turn it on at this time.

Whether a fetus remembers, prefers, and learns from his father's voice is not so clear. A team at the University of Toronto recorded mothers and fathers reading two minutes of *Bambi* and played the recording to fetuses. Only the mother's voice inspired a quickened fetal heart rate. That said, my baby's head seems to bob in my husband's direction whenever he speaks, as if she were a compass and he a magnetic pole. He likes that. But most of the time, fetuses don't seem to respond much to their dad's voice. If it's any consolation to men, it may be that the baby doesn't hear them as well. At 125 Hz, the male voice has a frequency that blends in with the internal noise—the beating heart, the gurgling stomach, and so on. Female voices (and some music) have an average frequency of 220 Hz, which penetrates the uterine walls better. I suspect my husband compensates with volume.

Some of the strong preferences that fetuses develop in the womb remain for weeks and even months after birth, and potentially much longer. Bass frequencies travel best through liquid and tissue (they're not so much heard as felt), and infants exposed in utero to low, loud, fast, energetic music strongly prefer it. Hearing the cheesy theme song to a soap opera they knew from the womb, newborn babies calm down and tune in, whereas babies who hadn't heard it before do not. A scary thought: Could babies be born programmed to relax to junk TV?

Yet the same can be said for classical music. Fetuses that bobbed in their amniotic fluid as they listened to the bassoon passage from *Peter and the Wolf* prefer it to other music they hear after birth. Not so of babies who didn't hear it in the womb. Could we really cultivate their musical tastes in utero? In one British study, moms instructed to listen to any song of their choice every day in their third trimester, from ragas to reggae to rap, gave birth to babies who remembered and preferred the same songs at one year old, as evidenced by how long they looked at the speakers. This happened even though the little rockers hadn't heard those ditties since before they were born.

Our fetuses are also learning the sounds of their mother tongue. In the third trimester, their cerebral cortex—the temporal and frontal lobes—tunes in to regular, habitual sounds, especially ones spoken by the mother. They're learning elements of tonal pitch, timbre, intensity, and rhythm. If my fetus heard Mandarin right now, she'd know it's not the language she hears me speak. When a group of Canadian researchers played a recording of an anonymous woman reading a story in English, stopping, and then resuming in English, the fetuses of English speakers showed no change in heart rate. Blah, blah, blah, they seemed bored. It was just familiar language spoken by a stranger. But when the speaker switched to Mandarin midway, their heartbeat sped up, suggesting they were paying attention and could perceive that the two languages are not the same. The same thing happened among Chinese babies when they heard their mother tongue stop and the speaker switch to English. After a baby is born, select neurons will strengthen to sounds of familiar languages, and other neurons will wither and die off. Around the time they become toddlers, they slowly begin to lose the ability to "hear" all the nuances of noise.

Babies (and perhaps third-trimester fetuses) are genetically programmed to home in on real language, not gibberish, and real music, not random noise. We know this because when experimenters play recordings of a mother's voice speaking backward, newborns have different brain activity than when the mom is speaking normally. Babies are processing sound in their left hemisphere only if it's intelligible language. Yes, they're really listening. An infant giggles when she hears her parents make funny noises because she knows they're just being silly.

Even cries are influenced by the mother tongue. Right out of the womb, German newborns cry in a downward fashion. Their wails descend in pitch because German, the language their mothers speak, is defined by a falling intonation. French newborns cry in a rising pitch, because that is how French is spoken.

"Little pitchers have big ears," goes the saying, warning adults to be careful about what they say within earshot of the children. I hope this isn't anatomically accurate. Especially now that I'm just weeks from giving birth.

> Music and rhythm find their way into the secret places of the soul.
>
> —Plato

WHAT CAN MOZART (OR ANY OTHER MUSIC) REALLY DO?

Given that fetuses hear and respond to sound, what might prenatal exposure to music do for them? Jeanne d'Albret, the mother of Henry IV of France, had a theory. She was convinced it shapes a baby's temperament. Pregnant with the future king, she called a musician to her chamber every morning to play a joyful, soothing tune. For the record, Henry IV "the Great" turned out to be an unusually jovial, kind, and beloved ruler.

Modern science has since confirmed that music can manipulate a fetus's mood. A little one's heart rate speeds up if he's interested or agitated by a sound and slows down when he's soothed. For instance, in an experiment at the London Maternity Hospital, Mozart proved to be a calming influence among the under-zero set. His sonatas especially seemed to pacify fetuses, slowing and steadying their tiny hearts, as did Vivaldi's, probably because many passages clock in at around fifty-five to seventy beats per minute, just like Mom's resting heartbeat. Rock music and raging passages of Beethoven and Brahms made fetuses restless. Some neuroscientists proposed that Mozartian musical sequences calm fetuses because they repeat themselves regularly every twenty to thirty seconds—just like the cycles of sleep.

Soothing was probably the original "Mozart effect," but the focus turned to smartening. It started in the mid-1980s when a music professor, Donald Shetler, played various types of classical music, including Mozart, to thirty pregnant abdomens. Observing the children after their birth, they appeared to him to have superior attention and language skills. At two years old, many of them could finger a simple melody on a piano. This was astonishing because most kids that age simply pound on the keys. Shetler's was a small research project and the findings were preliminary, but anyone who heard about his work was intrigued.

Then, in the early 1990s, Francis Rauscher and Gordon Shaw at the University of California at Irvine found that when volunteers listened to Mozart's *Sonata for Two Pianos in D Major* (and only Mozart, not any other composer), their performance on spatial reasoning tasks temporarily improved: they could more easily figure out ratios and manipulate objects in their minds. Spatial reasoning ability, enthusiasts argued, might translate into superior math and science skills, and everything else that requires higher brain function. Rauscher and Shaw claimed the music could boost IQ by as much as eight or nine points. But there was a catch: the subjects in the study were thirty-six college students, not infants or fetuses. Oh, and the spell lasted only about ten minutes.

Never mind the details. Almost immediately, the "Mozart effect" reverberated around Lamaze classes and preschools, amplified by anecdotal accounts. Add to the mix what appeared to be scientific proof—experiments on rat brains. A study, also led by Francis Rauscher, reported that rats exposed to Mozart's *Sonata for Two Pianos in D Major* while in the womb could complete a maze faster and with fewer errors than rats exposed to white noise or Philip Glass's modernist pieces. Another research group found that rat pups exposed to one hour a day of music before birth enjoyed an increase in new neurons in the hippocampus, resulting in better spatial learning ability, whereas rats exposed prenatally to random noise grew fewer new neurons and had impaired learning. Not all scientific studies achieved the same result, however.

By then, enthusiasm about the Mozart effect crescendoed into a mass media blitz and then a cottage industry of Mozart products: recordings, books, prenatal sound systems, classes, and so on. Marketers started pushing Mozart for everything that ails us, from prisoner unrest to roses that fail to thrive. It was starting to feel like a scientific legend, like the one about how French women don't get fat.

Eventually it was time for the Mozart effect to face the music. Some researchers criticized the practice of putting loudspeakers over a pregnant woman's abdomen because amniotic fluid amplifies certain frequencies that could hurt a fetus's eardrums. Where do we draw the line between stimulation and overstimulation? Other critics pointed out that apart from the anecdotes of boastful parents, there's little to no scientific proof that music-exposed fetuses have a head start—musically, mathematically, or spatially. It's unclear what part of the fetal brain is responding to music and whether the fetal temporal cortex is sufficiently mature for the music to stimulate it and improve spatial-relations ability. The death knell came when a Harvard researcher analyzed the net result of the sixteen studies on the Mozart effect and found it to be statistically insignificant.

This isn't to say that music is not beneficial for fetuses. I'm an optimist when it comes to this type of research. I believe that like many other things in life and in the lab, we need to know what we're measuring.

Which brings us to the coda of the original Mozart effect: music soothes and relaxes fetuses. This we know for sure. It's no small feat, for mood can indirectly affect behavior, including performance on intelligence tests. In pregnancy, your mood has a great impact on the fetus. Music influences your emotions, and your emotions shape your baby's mind. Does the fetus even need to benefit from the music directly? Turns out, no. One experiment found that when mothers put headphones on and listened to Mozart and other relaxing music, their fetuses relaxed and their heartbeats steadied, even though only the moms could hear the sound.

That's exactly the point. If listening to Mozart or new age music, humming a lullaby, or taking the time to focus on any music calms you, then you're producing calming hormones that are transferred to your baby through the placenta. Your baby's brain benefits from any buffer to the toxic effects of excess maternal stress. The music needn't be Mozart; anything that puts you in the right mood may work. One study found that among horror story lovers Stephen King's books have just as much positive impact on mental test performance as listening to Mozart.

Mozart himself sums it up best: "Neither a lofty degree of intelligence nor imagination nor both together go to the making of genius. Love, love, love, that is the soul of genius."

> You know you've got to exercise your brain
> just like your muscles.
>
> —Will Rogers

WILL EXERCISE STRENGTHEN BABY'S MIND?

Might prenatal exercise boost the muscle between my baby's ears? I hope so, for here I am at the gym, walking on a treadmill like a hamster on its wheel. My heart rate clocks in at a steady 140 beats per minute, the maximum threshold some doctors advise for pregnant women. Yet my pace is slow enough to read journal articles propped up against the treadmill's screen. So poky am I that the muscled gym boys on the machines to my left and right strike up a conversation over my head, as if I am only separating them for an instant as they pass by. I'm sweating as much as they are even though I'm moving at only a fraction of their pace.

It doesn't take much to make a pregnant woman sweat. Between the increased blood volume and jacked-up metabolism, blood rushes to the surface of our skin (think flushed cheeks). Expectant moms who exercise are more energetic and have healthier blood pressure, fewer varicose veins and puffy limbs, lower blood sugar level, a reduced risk of preeclampsia, and fewer aches and pains. We have less depression than if we were sedentary.

But enough about me. What's the baby getting out of my efforts? One article I'm reading gets my heart racing: "The Influence of Maternal Treadmill Running During Pregnancy on Short-Term Memory in Rat Pups." It's an experiment by Korean scientists on the effects of prenatal exercise on the brains of baby rats. Like me, the pregnant dams in the study run at a moderate pace on a treadmill for thirty minutes daily. We're all pregnant mammals on hamster wheels.

This is where the similarities end, because researchers couldn't do an experiment like this on humans. Three weeks after the exercising rats gave birth, scientists seized their babies and put them on a platform that faced a grid of stainless steel bars. By trial and error, the baby rats learned that they'd get a twenty-second electric shock if they stepped on the grid. One week later, the human equivalent of a few months, the researchers brought the babies back to the platform. The point was to see if the animals remembered the previous week's session. A long delay in stepping on the electrified grid would mean they recalled the shock and were trying to avoid it.

Who was more likely to remember the lesson: the babies of runners or the babies of sluggards? It turns out that Edison was right: genius *is* 99 percent perspiration. The babies of exercisers took longer to step off the platform onto the shock grid, suggesting they associated the grid with getting shocked. Their memory was better than their peers whose moms had been sedentary during pregnancy. They minded the zap.

Curious to see if the brains of these babies are any different or special, the researchers took tissue samples from their hippocampus, the region that controls memory. What they found was striking. Compared to the control group, the offspring of exercisers had a frenzy of new cells. Normally many cells die, but fewer did in the brains of rats whose mothers exercised in pregnancy. The more cells that hang around in the hippocampus, the more neural connections are made, and the stronger and speedier the memory.

Memory is the foundation of intelligence. How well we retain information and connect it to all else we know translates into smarts. A superior hippocampus increases memory powers, which in turn increases our access to everything we learn in life. By this measure, baby rats whose mothers exercise really are more intelligent.

This isn't the only study that suggests exercising mothers give birth to smarter babies. Others have sent expectant rats to spin around wheels, swim laps in rodent-sized Olympic swimming pools, and run on motorized treadmills. These jocks had babies that were better than others at navigating through mazes (superior spatial skills). One intriguing Canadian study even found that voluntary aerobic exercise among pregnant rats reversed the brain-damaging effects of prenatal alcohol exposure in their pups.

Do women who work out also give birth to smarter babies? There's much promise here. In one small-scale study on humans, newborns whose moms got aerobic exercise during pregnancy performed better on tests of alertness and stimuli tracking than did babies of nonexercisers. In another, the newborns of exercisers oriented and quieted themselves down after sound and light stimuli faster and more easily than their peers. There may even be real benefits over the long run. A small study at Case Western Reserve showed that five-year-olds of consistently and vigorously exercising moms had higher cognitive ability—better oral language skills and higher scores on tests of verbal comprehension, working memory, reasoning, and processing speed—than children of nonexercisers.

Prenatal exercise may directly affect a baby's developing brain by stimulating and preserving growth of neurons in the hippocampus, as seen in the babies of rodent jocks. Another theory is that working out may also elevate levels of a molecule called brain-derived neurotrophic factor (BDNF), which blocks the toxic effects of stress on a fetus's brain.

After thirty minutes of light-to-moderate treadmill exercise, I put my hand on my belly. The baby is as still as a sloth. I try to imagine my workout charging her mind. I have a vision of my energy transferring to her and becoming brainpower.

But her stillness, even when poked, reminds me of the downside of working out during pregnancy. Too much exercise has its drawbacks. Exercise is about mobilizing fats from storage to be used by muscles. Pregnancy is about storing and building up fat stores for the baby. When my heart rate rises, I am reducing blood and oxygen flow to the main uterine artery that nourishes the placenta. This results in a reduction in fetal body movements and an increase in fetal heart rate for about twenty minutes (although some studies show that the fetal heart rate may also slow down). Although I am sweating, my fetus cannot, so her little body begins to store heat. None of this is dangerous in moderation.

But in excess?

The closest we can come to addressing the question is by creating a boot camp for pregnant rats. At rat boot camp, expectant moms are forced to swim two hours daily, six days a week. Surprise: their babies don't turn out nearly as well as those born to mothers who exercised moderately. Boot camp moms bore pups with low birth weights. In one experiment, their mothers were so physically stressed that 19 percent died during lactation, and over 50 percent killed and ate their pups. This is never the desired outcome. And what's worse is what happened to grandchildren of those dams: they were born with low birth weights and grew slowly, even though their own parents had relaxed lives. The trauma echoed on through the generations, perhaps through a stress-related epigenetic effect (see more on this on page 137).

Boot camp is an extreme situation, but there may be a lesson here for women who exert themselves too much during pregnancy, whether by choice or circumstance. Exercise is a Goldilocks story. Too little is bad, and so is too much. The American College of Obstetricians and

Gynecologists recommends 30 minutes daily of moderate exercise such as brisk walking or swimming. But we each have our own thresholds. This is why pregnancy exercise advice is larded with caveats: don't pick up a high-intensity exercise if you did not do it before you got pregnant, be extra cautious if you have a high-risk condition such as an incompetent cervix or multiples, don't lose your breath, and avoid dangerous sports such as skiing or diving. Don't expect that working out will give you a shorter or longer gestation, or an ideal birth weight or Apgar score for your baby, for it has not been proven to do so. Fit women, however, may have a faster second stage of labor; that is, they can squeeze the baby out of the birth canal like a torpedo. The faster the baby leaves the birth canal, the lower the risk of oxygen deprivation and other brain-damaging effects.

I'll never know for sure whether my wimpy workouts really make my baby's memory stronger, but I like to think they do. Something about the gym—perhaps the two hairy men on treadmills attempting to outrun each other—reminds me that the world is so competitive. I'd love my daughter to have a head start at the starting line. But I know the best reason to work out is for the pregnancy to work out.

> Oaks grow strong in contrary winds and diamonds are made under pressure.
> —Peter Marshall, preacher

COULD FETUSES THRIVE ON STRESS?

For months I've wondered, if my baby is what I eat—all the chocolate cake and fish oil pills—isn't she also that which I feel? This would mean she's my anxiety about swine flu and congenital diseases. She's my rage at a loud, loutish neighbor. She's my fear about the delivery. She's my manic determination to finish a long article under a tight deadline. I had a heart-racing moment late in my second trimester when someone close to me was diagnosed with cancer. For hours my daughter was very still, like a mouse hiding in a hole. But that night, and over the following few days, she made a lot of distressed little kicks. I have no doubt that an unborn baby feels his mother's stress. I've stressed a lot about stressing out the baby.

But now I have relaxed a little. I'm calmed by scientific evidence that stress can be good for the fetus. We're so used to hearing that all anxiety and intensity in pregnancy is bad, so this is a relief. It turns out that psychological stress, like exercise, can be good when we strike the perfect balance—not too little and not too much. This is a good general rule. According to the Yerkes-Dodson law of stress optimization, when arousal levels are very low, we're understimulated. We're so serene that we're sleepy. In this state, we perform poorly on intellectual tasks and remember very little of what's going on. The same is true at the other extreme: when we're overaroused—too stressed—we also perform poorly and remember little. The sweet spot is right between the two. We're sharpest under some pressure, when we're in a novel or unpredictable situation, an unfamiliar environment or in competition, or otherwise not in complete control and yet not panicked. This is why athletes often perform better in the actual game than in practice. It's why we're often our most eloquent under some pressure. It's why some of us do our best work under a deadline.

Moderate stress is bearable stress. Say your life is moderately stressful, as it was for pregnant rats in a Japanese study on prenatal stress exposure. Neuroscientists at the Yamaguchi School of Medicine restrained expectant rodents in a small cylindrical steel cage for a half-hour daily in midpregnancy. For a rat, being caught in a cage is the human equivalent of your being stuck in a subway car going in an unexpected direction. Rats don't panic when trapped like this, but, like us, they really don't like the loss of control and uncertainty. Their stress levels may be the equivalent to ours when public speaking for a half-hour or driving without the GPS.

The moderately stressed rats went on to have their babies, which the scientists tested under various conditions. In a test of memory, the pups were trained to try to avoid a shock as they crossed a box. In a test of spatial learning, they were placed in an eight-armed maze. And in a third experiment to measure emotional reactivity, they were plopped in the center of an open field and observed.

The researchers were astonished. Compared to the control group, the rats whose moms were semistressed in their second trimester performed better on all three tests. They were quicker to remember and avoid shocks in the fear-avoidance test, and they navigated their way around the maze better. They were more wary and watchful when exposed in an open field.

What made rats with semistressed moms so much smarter, if not more cautious? The answer lies in the hypothalamus, which controls hormone production, and the hippocampus, the area involved in memory formation and storage. In small quantities, glucocorticoids (the most potent one being cortisol) are necessary for the development of these brain regions. Zapped by cortisol, our senses and memory perk up, as they do when we're under pressure to, say, give a presentation, or when we nearly crash into the car in front of us when driving. We think more clearly, we're more alert. When these baby rats were fetuses in their mother's womb, their tiny brains were exposed to shots of cortisol and other glucocorticoids. These hormones fertilize neuronal growth in the hypothalamus and hippocampus after a bout of stress much in the way they do after a round of exercise. Both moderate physical stress and moderate psychological stress do a fetus good.

The idea that moderate prenatal stress may be good for a baby is startling enough. Even more provocative is the possibility that so may *pre*-prenatal stress—how stimulating and positively stressful our environments were *even before we got pregnant*. A fascinating study led by biochemist Larry Feig at Tufts University found that if you put young prepregnant rats with genetic memory problems in an enriched environment—with active social lives, exercise wheels, and novel toys—their capacity to remember improves. This is remarkable by itself. But what astonished the researchers is that the *babies* of those rats also had an improved capacity to remember, even though those babies were conceived after the moms were removed from the enriched environment, and they themselves grew up in a nonstimulating setting. The mechanism that fixed the mothers' genes or compensated for them may have been passed on to their offspring—another triumph of epigenetics, the environment affecting the behavior of genes. Might a stimulating, challenging lifestyle make an impact (on our eggs) before conception as well as (the fetus) during pregnancy?

There is already some evidence that what's good for dams may indeed be good for us dames and our babies too. Janet DiPietro, a developmental psychologist at Johns Hopkins University, has led research that found that moderate prenatal stress in mid- to late pregnancy is associated with higher cognitive and motor scores in children. Compared to women who had a somewhat carefree nine months, moms who reported moderate anxieties and pressures during their second and third trimesters gave birth to babies with faster neural conduction for three of the four tests of brainstem auditory-evoked potential, an indicator of a more mature brain and nervous system at two weeks old. These are women who felt stressed about the usual everyday stuff: their ability to do tasks and chores, their weight, getting enough sleep, whether the baby would be normal, and so on.

In an earlier study that tracked babies for longer, until age two, DiPietro and her colleagues found that prenatal stress and even mild depression midway through the pregnancy were associated with a child's higher scores on standardized tests of motor and mental development. In a study by a different group at the University of California at Irvine, one-year-olds whose mothers had elevated levels of cortisol in the second and third trimesters (though not reported maternal anxiety) scored higher on cognitive tests.

High-scoring babies didn't always have the sunniest temperaments, however. Prenatal stress is also associated with increased infant irritability. DiPietro warns us not to confuse a baby's information processing ability and brain maturity with temperament.

There's an important lesson here. We shouldn't worry that we're harming the baby when we feel somewhat overwhelmed, off-kilter, or out of control. Maybe we should even welcome the stress. Bring it on. We're learning from these ups and downs, and so are our babies-to-be.

But what happens if we move the stress dial from medium to high?

———

In his 1981 album *Look How Me Sexy*, the Jamaican musician Linval Thompson wrote a song, "Baby Mother," that calls for men not to stress out their pregnant companions. "Mind how you're pushing/when you push on your baby mother," he sings, warning men that when they threaten a pregnant woman, they hurt the unborn baby too.

Thompson was on to something. If a pregnant woman's partner is chronically cruel, leaves her, or dies while she's pregnant, the child is more likely to suffer from depression, anxiety, attention-deficit disorder, and criminality in adulthood. This is extreme stress: going through a divorce or separation; being a victim of domestic violence, crime, or discrimination; moving, changing jobs, or suffering from the loss of someone close to you; and being physically ill. Kids whose moms were overly anxious and feeling a loss of control during pregnancy—for any reason—have been reported to be more irritable, depressed, and clingy; have sleeping and feeding problems; have emotional problems; and get lower grades in school.

Natural disasters, too, are a cause of extreme prenatal stress, and they leave their mark on unborn babies. Studies of birthrates after severe earthquakes in China and New Zealand have found that significantly more women go into preterm labor immediately afterward. Extreme stress and depression make our bodies want to evacuate; it's all about self-survival in an emergency. The long-term prognosis isn't rosy either. A study of survivors of an 8.0 Richter scale earthquake in China found that eighteen-year-olds who were in utero at the time of the disaster had lower IQ scores and were more depressed than peers and siblings who were born just one year later.

Similarly sad results come from Canada, where a devastating ice storm hit in 1998. Children born to moms who suffered high levels of stress during the disaster—imagine forty days of power outages or living in a shelter during the coldest months of the year while pregnant—scored lower on tests of full-scale intelligence and language abilities, including speaking and understanding fewer words, when tested at two and a half and five years old.

If studies on extreme stress in pregnant rats are any indication, what may have happened to these kids is that Mom's steady drip of stress—or one great tidal wave of it—deluged the placenta, which normally breaks down glucocorticoids (cortisol) to fetus-friendly doses. Overwhelmed, the placenta allowed toxic levels of cortisol to pass through like a busted dam. In moderation, cortisol helps neurons in the hippocampus, the memory bank of the brain, to grow. But fertilizer in excess is poison. It's toxic, and it terrifies the fetus. The fetus effectively absorbs the anxiety, resulting in learning and memory problems.

(Licorice lovers beware: The treat may help cortisol sneak through the placental barrier. It contains the compound glycyrrhiza, which inhibits an enzyme that breaks down cortisol, thereby raising fetal glucocorticoid brain levels. According to a Finnish study, licorice lovers have babies that score lower on cognitive tests and have problems with attention and aggression—similar to kids exposed to extreme prenatal stress. All this suggests that we may want to limit the licorice.)

Cortisol-poisoned, a fetus may become a jittery baby who could grow up to be an anxious adult. Excess glucocorticoids impair the development of the baby's hypothalamus-pituitary-adrenal axis—the system that controls the stress response—programming it to react too often and too easily, like a smoke alarm that wails when there's just a whiff of burned toast in the air. Over time, the stress causes high blood pressure, diabetes, asthma, and a weakened immune system—diseases commonly found in the adult children of moms who were overly stressed during pregnancy. The amygdala, the emotional region of the brain, and the prefrontal cortex are also battered. This leads to anxiety, sleep disturbances, misdirected attention, and other high-strung behaviors.

Your stress hormones during pregnancy give your unborn baby an idea of what kind of world he will arrive in. If your system is flooded with prenatal glucocorticoids, your baby gets the hint. His body prepares for a life of stress. From an evolutionary perspective, a useful response to living on the edge is to be anxious and high-strung. Sure, a person growing up like this might be depressed or have attention deficit disorder in a calm setting. Given a proper emergency, she's hyperfocused, guarded, wired, and hypervigilant. We all know these people.

They're shock jockeys, adrenaline freaks, or worrywarts. They might have high blood pressure or heart attacks if they live long enough, but in a pinch, all that matters is survival. It's as if being exposed to such conditions prenatally programs babies for shorter, nastier, more anxious lives. That's the trade-off for heightened vigilance.

In the absence of real danger, you'd probably wish for your baby to be exposed to just enough cortisol to give her an edge. You'd want her to be aware and sensitive to her surroundings but not overstimulated by them. You'd want her temperament to be set for a competitive, spirited world—not a panicked, apocalyptic one.

Of course, there is no one-size-fits-all-stress scenario. This makes the effects of prenatal stress very hard to predict. What's excruciatingly stressful for me may be only mildly diverting for you. Stressed-out kids may be born to stressed-out moms because they're genetically predisposed to be stressed out. (Some genes have been found to confer stress-resistance, while others are linked to nervous dispositions.) Growing up in a troubled environment might trigger vulnerable genes.

The timing of prenatal stress matters too. We care about the fetus's gestational age when exposed to cortisol because the brain matures in stages, and each stage of development has a different window of vulnerability. Some studies indicate that the fetus's brain is most vulnerable to the negative effects of excessive stress hormones in the middle of the second through the third trimester (yet it benefits the most from moderate, or "healthy," stress at this time).

Male and female fetuses respond to stress differently. Male brains develop more slowly than female brains, making them susceptible to stress hormones for a longer period of time. This may be why boys exposed to overwhelming prenatal stress are more likely to suffer from learning deficits while girls become anxious and depressed. Male fetuses are also more likely to thrash around when Mom is freaking out. They are more vulnerable to nervous tension, and they are more likely than females to miscarry.

Even after taking into account these findings, it's hard for scientists to pin down why some children seem more affected than others by extreme prenatal stress, even when their moms were equally stressed.

This is where the most important lesson about prenatal stress comes in.

_____

Not long ago, a group of researchers at Rochester Medical Center in New York recruited 125 women in their second trimester of pregnancy and gave them a standardized questionnaire to measure anxiety and stress. Then they took a sample of each woman's amniotic fluid and measured it for the stress hormone cortisol. Between the two tests, they got a pretty clear picture of their volunteers' stress profiles. About seventeen months after the women delivered their babies, the researchers tested the toddlers for the effects of prenatal stress on their development.

Since the researchers knew the mothers' stress levels during pregnancy, they were able to look for a connection between stress and development. They used a standardized developmental scale test to assess the toddlers' attention spans, play activity, puzzle-solving skills, and so on. Then they tested the children's bond with their mothers using standardized measures of secure or insecure attachment. The mothers left the room, leaving their little ones with strangers. Was the toddler unable to separate from her mom? (Insecure.) Did he seek and receive comfort from Mom when frightened? (Secure.) Was she happy to see Mom when she returned? (Secure.)

The results were inspiring and relieving. If the babies clearly had secure, loving relationships with their moms, the link between high prenatal stress hormones and impaired cognitive development was eliminated. A mom may have been a bit of a nervous wreck during pregnancy, but her child's nervous system was spared. Not so for kids who had troubled relationships with their moms. (See "Is Maternal Instinct Shaped in Infancy?" page 195, for further insights.)

This is tremendously hopeful news. It suggests that the environment after birth may help reverse some or all of the damage of prenatal stress. A predisposition for fear and overanxiety in an insecure kid may become something more like focus, conscientiousness, and introspection in a secure kid. Predisposition doesn't mean inevitability. Tendency isn't destiny. If, like a ticking bomb, a baby had been wired for stress in the womb, then nurturing may defuse the bomb.

How Fetuses Calm Us

If you were to take the Trier Social Stress Test, as nearly 150 pregnant women did in a study at the University of California at Irvine, you'd be led to a windowless room with a video camera. There, an assistant would ask you to sit while he or she hooked you up to instruments that measure vital signs. You'd also encounter three men and women sitting at a table, waiting for you.

They'd be your interview panel.

Facing them belly-on, you're to pretend you're applying for a job and must deliver a five-minute speech to convince the panel you're right for the position. And when you're done, you'll be asked to do a bit of mental math—say, to count down, in increments of thirteen, from a large prime number like 313. Before and after the ordeal, a researcher will enter the room to collect your saliva for stress-hormone testing.

Analyzing all the data from the stress test, including the women's body language and hormone levels, the researchers confirmed something remarkable: the further along a mom was in her pregnancy, the less stressful she found the stress test. Compared to their stress levels in second trimester (seventeen weeks), volunteers in their third trimester (thirty-one weeks) had lower blood pressure, a slower heart rate, and a lower rise in the hormone cortisol. Nonpregnant women who took the same tests at the same time intervals stressed much more. This was not the first study that found that pregnant women, especially those in their third trimester, are calmer than nonpregnant women under the same somewhat stressful circumstances.

So what is it that makes pregnant women so much more Zen as we approach our due dates? The answer is that the body reduces the sensitivity of its cortisol receptors even though it's producing higher levels of cortisol. This means it takes more cortisol than usual to get the nervous system all hot and bothered. At the same time, the placenta produces more of an enzyme that changes cortisol to an inactive form. Less toxic stuff reaches the baby. As we close in on the third trimester we also produce more soothing oxytocin and prolactin. Like a nightcap at the end of a long, tough day, this hormonal cocktail helps us relax and bond.

All this is good news for moms who are slammed with short-term mild to moderate stress late in their pregnancies. But there's an even bigger surprise to come out of this. We may think of this as our body unconsciously protecting the baby at a time of stress, but it's just as valid to think of it the other way around: our babies are protecting us (as well as themselves). After all, it's their placentas that are dampening our response to cortisol and making us more stress resistant. It's like yin-and-yang: mother calms baby, and baby calms mother.

> Stress wouldn't be so hard to take if it were chocolate covered.
>
> —Anonymous

DO CHOCOLATE LOVERS HAVE SWEETER BABIES?

I admit it: I turn to food when I'm stressed. One reliable soother is chocolate. Chocolate gets me in trouble with the pregnancy police. All the sugar and fat! Not to mention the caffeine!

So imagine my delight when, not long ago, the stunning super-model Heidi Klum, well into her third trimester with her fourth child, presented a photo of her pregnant self smothered in chocolate. She did this on the *Ellen DeGeneres Show,* in front of a live audience. The crowd gasped when they saw the pregnant Klum looking like a chocolate-covered strawberry.

"We were doing our normal shoot," explained Klum. "I said I would love to be covered in chocolate. So someone ran out to the drugstore and got, I think, seven or eight of those squirty chocolate things, and then they just squirted over me and did the photo." She put on a happy, innocent look. "It was yummy!"

"I bet! I bet it was yummy," Ellen said agreeably. She opened her eyes very wide.

"It was perfect. Especially when you're pregnant!" grinned the supermodel. With an immaculately lipsticked mouth, she mimed what she thought sloppy gobbling might look like.

The women in the audience cheered and clapped.

We celebrate the implicit symbolism of an expectant mom, her naked skin plump and radiant and immersed in primordial ooze as rich, dark, and fertile as mud. Chocolate is one of the treats pregnant women crave most, a fetish food. All is bared here: sex, lust, fertility, exuberance, and indulgence. Even the word *chocolate* is symbolic. It's associated with Xochiquetzal, the Aztec goddess of love, fertility, and pregnancy.

———

But can chocolate actually be good for the baby? The most interesting research on its effect on fetuses took place several years ago, in candy-loving Finland, by Katri Räikkönen. A professor of psychology, Räikkönen decided to conduct a study on how a pregnant woman's consumption of chocolate might affect the disposition of her baby-to-be. Räikkönen and a team of colleagues, including a neonatal pediatrician and medical doctors, recruited over three hundred women in their late twenties and early thirties who had just given birth and were recovering in the maternity ward. The researchers knew from previous studies that pregnant women who are stressed tend to eat chocolate to soothe themselves. They also knew that frazzled women tend to give birth to the most sour and irritable babies. Is it possible, they wondered, that of all the newborns in the nursery, those born to chocolate-loving mothers would have better temperaments?

To find out, the researchers asked the new moms to report how much and how often they ate chocolate in the previous nine months, on a scale from never to weekly to daily, and to mark their level of psychological stress on a visual scale ranging from none to maximal. The women gamely answered the questions and took their babies home. After about six months, when they likely forgot ever having participated in the chocolate consumption half of the study, each of the moms received a survey in the mail asking her to rate her infant's temperament and disposition. The researchers wanted to know: How often did her baby laugh? How easily was her baby soothed? Did the baby respond well to novelty? How active was he? How easily frustrated was she?

It is striking how chocolate consumption correlated with infant temperament and activity. Chocoholics, the type who indulge every day, did indeed have babies described as more outgoing, less fearful, less frustrated, and more easily soothed than their counterparts who ate chocolate weekly, seldom, or never while pregnant. Their infants smiled and laughed more than the other babies. They were, well, sweeter. Weekly chocolate eaters, the majority of the women, had infants who were somewhat more active and reactive than those whose moms completely shunned the treat. It's safe to say there was a clear trend: the more chocolate a woman consumed in pregnancy, the sweeter her baby's disposition.

Räikkönen and her colleagues went further, delving into the stress response. As expected, they found that the more stressed a woman was when pregnant, the more fearful her baby turned out to be (generally). But what's astonishing is that chocolate appeared to modulate fright and distress ratings. The team zeroed in on the most anxious, timid infants and found that they were much more likely to have moms who seldom or never ate chocolate. When a mom reported eating chocolate weekly or daily during pregnancy, her stress level did not adversely affect her baby as much as those whose moms shunned chocolate. It's as if there's something in chocolate that somehow protects against— or even reverses—the negative effects of excessive prenatal stress.

So, what's in the secret sauce (or syrup)? Well, the researchers themselves would insert a caveat here. Correlation is not causation: we can't be sure chocolate gets all the credit. The researchers fudged the difference between dark and milk chocolate, and they relied on self-report. Are babies of chocoholics really sweeter, or do these moms, in their sweet chocolate high, merely perceive their babies as such? It's a valid question, although studies have not found any association between chocolate consumption and any personality trait, including optimism. (Yes, experts have studied "chocolatist" personalities.) There's always a possibility the "sweet baby" effect could be postnatal, boosted by chocolate chemicals in the breast milk.

But the most tempting theory is very real and possible. That is, one or more biochemically active compounds in chocolate might really affect the behavioral characteristics and temperament of infants. It might make them sweeter, beginning in the womb.

An obvious reason why chocolate lovers have sweeter babies is that chocolate is uplifting and stress-relieving for the mom. If chocolate calms your nerves, you're producing fewer stress hormones that could adversely affect your unborn baby's moods and temperament. You're effectively short-circuiting your stress response. Chocolate's stress-relief properties lay in the opioid reward system, and pregnant women may be especially susceptible. The sugar-and-cream rush-and-lull of chocolate releases feel-good endorphin opioids in the hypothalamus, enhancing the activity of dopamine, a neurotransmitter that improves mood. Tryptophan, an amino acid present in scant amounts, may make you feel content and drowsy.

The Finnish researchers have other theories too. One is that babies of chocoholics are less stressed because they were exposed to phenylethylamine in the womb. This psychoactive stimulant inspires the "chocolate theory of love" because it triggers the release of dopamine. While very little phenylethylamine actually makes it to an adult's brain before being metabolized, it may cross the placenta and affect the fetus's brain, which is much more sensitive. While an adult would need to eat more than twenty-five pounds of chocolate to experience a marijuana-like effect, the fetus obviously can make do with less.

Speaking of pot, chocolate also contains a potent cannibinoid called anandamide. Anandamide enhances a sense of pleasure and well-being by activating the same cellular receptor compounds found in marijuana. (The word *anandamide* literally means "internal bliss.") Other candidates are anandamide's cannibinoid cousins (N-oleoylethanolamine and N-linoleoylethanolamine). Also found in the average chocolate bar, these compounds are known to enhance and prolong pleasure by delaying the breakdown of the anandamide we produce in our brains naturally.

It's easy to believe that a feel-good fetus—or at least one buffered from the toxic effects of excess stress—would be a happier infant and, who knows, maybe—because these early days are a foundation for life—a happier adult.

The fact that so many chemicals in chocolate can sneak across the placental barrier is interesting. Theobromine, a caffeine-like stimulant found in dark chocolate, is another pregnancy-friendly compound. Researchers at Yale asked over two thousand expectant moms about their chocolate consumption in pregnancy and measured theobromine in their cord blood at delivery. The chocoholics had the highest amounts. Women who ate five or more servings of chocolate weekly in their third trimester, especially of dark chocolate, were nearly 70 percent less likely to get preeclampsia. Another study found a slight but significant decrease in miscarriage rates in the first trimester among daily chocolate eaters. Researchers credit the stimulating effect that theobromine has on placental circulation while blocking oxidative stress.

"Excuses, excuses!" my skinny friends chide when my pregnant self expands on the reasons to indulge. I'm obliged to add that moderation is warranted here. The moderate mom-to-be should limit her chocolate consumption for two reasons: caffeine and sugar. If I limit my caffeine intake to two hundred milligrams daily in pregnancy, the amount recommended by health practitioners, then I can gorge on about ten ounces of dark chocolate (about twenty milligrams of caffeine per ounce) daily, assuming it's my only caffeine source. Milk chocolate (about six milligrams of caffeine per ounce) has a lot of fat and sugar, so I try to watch that too, in a largely symbolic effort to ward off gestational diabetes. A daily dark chocolate bar seems temperate—and tempting enough.

I must confess: after I eat a chocolate bar, my baby really does seem livelier. She bobs and kicks in my belly. She dances happily as I play Mozart and Tom Waits. I melt. "I have the sweetest kids," chocoholic Heidi Klum bragged. Well, *I* have the sweetest fetus.

Or maybe that's just the chocolate talking.

Estimated amount of glucose used by an
adult human brain each day, expressed
in M&Ms: 250

—Harper's Index

DO FIDGETY FETUSES BECOME FEISTY KIDS?

I eat chocolate, and the baby kicks. I watch a horror film, and the baby kicks. She kicks when I'm scared, and she kicks when I'm happy. For every one of my actions there is a reaction, and her reaction is to kick. She kicked my cervix anxiously when I watched a YouTube video of a C-section, as if she knew it augured her exile from Eden. Last week, I got mad at my friend's boyfriend—he had betrayed her—and the baby started punching and jabbing me in both the cervix and ribs. "How cute!" my friend said when I told her my little one got angry on her behalf. "My chihuahua does that, too."

Oh no. Do highly reactive fetuses become the chihuahuas of the under-zero set—hyperactive, oversensitive, irritable, and anxious?

It's actually a loaded question.

One of the research groups exploring fetal predictors of temperament is a team from Johns Hopkins led by Janet DiPietro, the developmental psychologist whose research suggests that mild to moderate prenatal stress is good for the baby. DiPietro asked volunteers who were thirty-two weeks pregnant to watch a half-hour of a documentary on labor and delivery that, like the graphic flick I saw on YouTube, was well intended but gory. Each mother's heart rate was measured, and so was her fetus's, the latter on a machine that also detected his or her every movement in the womb.

Watching the scenes of their peers going into labor and delivering their babies, pregnant women unconsciously breathed more heavily and sweated slightly. Sometimes their fetuses became more active with accelerated heart rates. More often, they got really still, and their fetal heart rates slowed down, as did their mother's. Surprisingly, many babies did not react at all.

When asked how they felt about watching the birth videos, most of the heavily pregnant women said: "Heartwarming! Uplifting! Enjoyable!" Who were they kidding? And yet women who had never given birth before, like myself, had fetuses with faster-thumping hearts and increased movement, while fetuses of experienced moms moved around less. It's known that women who feel stressed tend to have more fidgety fetuses, especially if they're carrying a boy.

The researchers were curious to know if the most reactive fetuses would become the most reactive newborns. Six weeks after giving birth, the new moms dutifully brought their newborns back to Johns Hopkins for a series of harmless tests. Researchers dandled the babies. They held them upright and laid them on their backs and their fronts. They undressed them, weighed them, and redressed them. Observers as watchful as helicopter mothers coded every grimace and giggle, scream and smile.

The researchers found what you might expect: the more fidgety a fetus was when the mother was stimulated in pregnancy, and the faster his heartbeat, the more excitable (e.g., irritable) he was after birth. These infants were more likely to howl and fuss and arch their backs when handled. But this was true in the other direction too: fetuses that froze and had slower-than-average heartbeats when mom was stimulated were also more irritable as infants than were the unfazed fetuses. The more extreme the response was in the womb, the more sensitive and reactive the baby would be later on. According to DiPietro, "Fetal motor activity may best be considered as an indicator of temperament"—how energetic, shy, fussy, adaptable, intense, or cheerful our babies will be.

Much of a fetus's behavior in the womb comes from the mother's response to her environment, which makes it an imperfect indicator of his individual temperament. So what happens if we try to stimulate the fetus without bothering the mom? This is what the Johns Hopkins group did in a small, earlier experiment on women between thirty-two and thirty-six weeks gestation. They pressed a vibrator-like device against women's abdomens. Fetuses that moved around a lot in response to the vibrations, or had high baseline heart rates, generally turned out to be the fussiest, feistiest, most active infants later when they were six months old. These babies didn't have regular sleeping or eating habits and responded poorly to strangers. (I know these vibrating wands. Late in pregnancy I took a stress test, which involves prodding the baby with one of them. My little one refused to budge until she heard her daddy walk in the room and start talking.)

Even in the same womb, one baby may act differently from another—a testament to individual temperament showing up early. When twin pairs in early second trimester were observed in an Israeli study, the more active of the two fetuses became the more unpredictable, difficult, and active baby at three and six months old. In a study that took a longer view, two-year-old babies who had been more active fetuses were more likely to embrace novelty—to interact with toys and strangers naturally and without fussing.

So here's the next question: If a fetus is excitable (or frozen in fear) the first time she experiences new stimuli, like a vibrating wand, will she be equally affected the second, third, or fifth time she's wanded? The answer is interesting because it turns out that how quickly a fetus habituates to stimuli might have something to do with development as well as temperament. A study at the University of North Carolina at Charlotte found that babies tested to have more advanced mental development at six months were once fetuses that adapted quickly to novelty. Remembering a stimulus—and perhaps knowing it's not really dangerous—is considered an early form of information processing.

Perhaps I'm projecting, as hopeful expectant moms do, but I wonder if this could be true of my baby and moxa, an herb I've smoldered near my pinky toe every night. Moxa is meant to startle and stimulate a baby into turning from breech position to a birth-ready head-down presentation. The first time I smoked it, she kicked and kicked. But after that, I could smoke the moxa until I was bug-eyed and choking, and she wouldn't flinch. Is this a sign of intelligence or stubbornness—or did I knock her out?

In the world of prenatal prediction, a fetus's resting heart rate may also predict developmental outcome. (The range in the third trimester is 112 to 165 beats per minute.) A lower resting heart rate is predictive of less crying and fussing in infancy. Monitoring fetal heart rate for nearly an hour at a time, the Johns Hopkins group discovered another telling pattern: fetuses with slower and more variable heart rates at twenty-eight weeks gestation and after scored higher on mental, motor, and language development tests at age two and two-and-a-half years than did fetuses with faster and more stable heart rates.

At the slower end of the normal range, and with more variability, is how you want your fetus's heart to beat. Heart rate, or cardiac patterning as scientists call it, is an indicator of a maturing nervous system, switching back and forth from automatic controls to higher cortical processes that pick up on nuances in the environment. Heart rate patterns tend to remain constant before and after birth, and possibly well into childhood.

Interestingly, stress is related to variability in fetuses' heart rates. When DiPietro and her team challenged pregnant women to take a word-and-color association test under time pressure, the women's twenty-four-to-thirty-six-week-old fetuses had more variable heart rates, a likely response to their moms' racing heart, higher blood pressure, and other signs of minor stress. (See "Could Fetuses Thrive on Stress?" page 133.) This is a good thing, the fetus learning about the world from her mother's bodily cues. Extreme stress, however, is associated with a less variable fetal heart rate.

Reading these studies, I've felt the need to grab the pocket Doppler and measure my baby's heart rate. I've held myself back, but just barely. Do I want to be one of those nerdy, overambitious parents, armed with gadgets, trying to predict their baby's temperament and intelligence long before birth? Of course—not. I know that all that poking and prodding, and my anxiety about the outcome and what it could mean, probably wouldn't give my little one the sweetest disposition.

> Clouds, leaves, soil, and wind all offer themselves as signals of changes in the weather. However, not all the storms of life can be predicted.
>
> —David Petersen, author

WHAT DOES BABY'S BIRTH SEASON PREDICT?

Our baby is due in July. It wasn't my choice to be heavily pregnant in high summer. Back when we had been trying to conceive, I had been hoping for a birth in the fall and my husband was rooting for spring. I was born at the bitter end of November, which naturally makes autumn my favorite season. My husband was born in May, when buds are bursting and skies are blue. I enjoy a melancholy savor, whereas Peter likes happy endings. My idea of release is to travel to a foreign country, with no schedule and no knowledge of the local language, meander, and get lost.

Could it be that our respective birth months, May and November, representing the opposite solstices, affect our personalities and behaviors? What would this mean for our summer baby?

This is not an astrological question. But it does involve the star we're all born under: the sun. For decades, scientists have been fascinated by the idea that seasonal fluctuations—the hours of daylight changing—may affect the fetal brain even in the hot liquid perma-dark of the womb.

Astonishingly, there may be something to it. Studies have found that young fall-borns and winter-borns (with birthdays from October to March in the Northern Hemisphere) are more sensation seeking than are spring-borns and summer-borns. If a baby is born in these months, she's likely to score higher on the SSS, the Sensation Seeking Scale. Chances are, her thrill-seeking phase will peak when she's twenty-six to thirty years old. Yes, these Ernest Shackletons (February), Yuri Gagarins (March), and Alexandra David-Neels (October), are more likely to be bored by other people's home movies, and they don't like it when they can predict what happens next in a film or a novel. They don't mind traveling to a new location without a timetable and are eager to explore and experiment. When dating, prospects who are "physically exciting" are preferable, and they appreciate "earthy body odors." Aesthetically they often prefer the clashing colors and jagged chaos of modern art to the clarity and harmony of the classical. They crave intensity.

At least, when they're young. Several studies suggest that by their thirties and forties, women born in the fall and winter reverse this rock-and-roll behavior and become less sensation seeking than their peers. Compared to people born in the rest of the year, babies born from October to March have a greater rise in novelty seeking during adolescence but a steeper plummet in middle age.

Mood has a lot to do with intensity and thrill seeking, and moods are modulated by dopamine. This neurotransmitter has a voice in whether we're extroverted or introverted, high-spirited or mellow, a busybody or a homebody. Dopamine zeroes in on the "emotional brain," or limbic system, which leads to the pursuit of pleasure and reward—food, drink, sex, praise, and so on. It's what moves us and shakes us, literally. When dopamine is depleted, we lose motor control, as is the case for Parkinson's patients.

Testing the blood of infants and young adults, scientists found the highest levels of dopamine metabolites among those born in November and December (me) and the lowest levels for those born in May and June (husband). (The months are reversed in the Southern Hemisphere.) This suggests that the increase in thrill seeking among people born in fall and winter may be related to the steeper increase in dopamine activity. Just as their interest in novelty declines more steeply later in adulthood, it's possible that their dopamine levels do too. Perhaps this helps explain why people born in the winter tend to be at a higher risk for bipolar depression and schizophrenia and score lower on tests of agreeableness.

What happened? Prenatal viral infections, pesticide exposure, vitamin D, and nutrition could have something to do with season-of-birth effects. But a leading explanation for why teens and young adult fall-borns and winter-borns have more dopamine activity than people born at any other time of the year involves their master clock, whose function depends on how much sunlight their mothers were exposed to in pregnancy or how much they got soon after birth. No one knows exactly how or when a fetus's body clock is wound and set, which genes are silenced and which ones are activated, but it likely involves the hormone melatonin, which can sneak through the placenta and bathe the fetal brain. Melatonin levels depend on light exposure and are highest at night and in the darkest months of the year. Your melatonin levels tell your fetus the time of day and season of the year, and your fetus uses this information to set her own rhythms.

In the darker months when melatonin levels are highest, dopamine is lowest. In the sunnier months with longer days, dopamine rises. For fall- and winter-borns, the clock setting might happen at a phase in pregnancy when days are getting longer. This is when pregnant women are exposed to more light and produce less melatonin. Does this signal the fetus to increase dopamine activity? It's unknown, yet once the baby's master clock is set, it keeps ticking long after birth.

Some researchers think a prenatally set master clock is also the reason that fall and winter babies are more likely to become early birds, the type of people who wake up bright-eyed and chipper around dawn. If nightfall arrives sooner and sooner around the time of the baby's birth, his or her circadian rhythm may favor a "phase advance," meaning a backward shift from, say, a 7:00 a.m. wake-up to a 5:00 a.m. wake-up. According to one theory, it's easier for fall- and winter-borns to adjust their circadian rhythms to the long days of strong sunshine than for spring- and summer-borns to adjust themselves to the feeble light of winter. Men born between April and September are more likely to suffer from seasonal affective disorder, a slump that strikes in late fall and winter.

Biological clocks, too, seem to be affected by the season in which we are born. Intriguingly, an Italian study found that if a baby is born in autumn, it's more likely that she'll be fertile later in life than a woman born in spring. On average, fall-borns reach menopause eighteen months later than do spring-borns (at age forty-eight versus fifty years). Prenatal factors possibly related to prenatal sunlight exposure, temperature, or diet may alter placental blood flow at certain phases and therefore the number of eggs a baby's ovaries contain at birth. The eggs we're born with are all we'll have in our lifetimes.

Another interesting finding is that the season of birth is linked to how lucky one feels in life. In a survey of over twenty-two thousand adults, most in their thirties, those who considered themselves luckiest were born in spring. These spring chickens, particularly the May babies, were more optimistic and extroverted—well, sunnier. They were more likely to expect good things to happen to them. Those who considered themselves the least lucky were the brooding November babies. (Incidentally, men born in autumn had fewer children and a higher probability of being childless, according to another study.) The researchers suggest there could be a dopamine connection here too. Perhaps these cheery spring-borns avoid the steep decline in dopamine activity that fall- and winter-borns get when they approach middle age.

There may even be an evolutionary reason that personality traits are affected by birth season, according to anthropologist Dan Eisenberg. Nearly twelve thousand years ago, the climate underwent phases when the earth was colder and darker for decades to centuries at a time, comparable to a semipermanent fall or winter. Even spring- and summer-borns were thrust into a world of diminished daylight. During these spells of decreased food availability, warfare, and low life expectancy (most people didn't make it into their thirties), certain behaviors could have been beneficial to survival—sensation seeking among them. It's possible that low light exposure during a window of development triggers restless, reward-seeking behaviors even if they come at the expense of overall happiness.

Studies on the link between season of birth and behavior are striking. Yet Jayanti Chotai, the psychiatrist who conducted many of them, warns us that the magnitude of the difference between seasons is so small that we can't predict how an individual baby will develop. These patterns show up only on a large scale. Other factors clearly influence personality more.

As for my July baby, I don't worry that her life will be dull. Her summer-born role models are plenty spunky: astronaut Neil Armstrong, explorer-entrepreneur Richard Branson, and the Dalai Lama among them.

> We continue to shape our personality
> all our life.
>
> —Albert Camus

WHAT CAN WE FORECAST FROM A FETUS'S FINGERS?

There is one form of scientifically endorsed fortune-telling. It's not palmistry or astrology, but it's close. It's called digit (finger)-ratio research, and so fascinated by it am I that I want my fetus's digits measured. And so, during my third-trimester ultrasound, I ask the technician to zero in on them. The hands, my baby's hands, stand out as a landmark in an otherwise ghostly, featureless landscape.

"Her hands?" the tech asks.

"The fingers," I say. "Please." She shrugs grumpily and zooms in on a perfect miniature appendage. The fingers are wiggling slightly like tentacles in the amniotic fishbowl.

I contemplate these fingers solemnly. The ring and index look about even.

Peter, who is sitting at my side, lets out a harrumphing little laugh. He knows what this is all about. I have been researching how levels of prenatal hormones in the first trimester of pregnancy, particularly testosterone and estrogen, affect the relative length of the index and ring finger.

My inspiration comes from John Manning, an evolutionary psychologist and professor emeritus at the University of Liverpool. Manning is the father of digit ratio forecasts. Back in the 1990s, he startled the science world by popularizing the very odd observation that the growth of the fingers is related to prenatal testosterone level, which in turn is linked to various personality traits.

As early as eight weeks into a pregnancy, Manning claims, sex hormones, particularly testosterone, work their charms on the brains and bodies of fetuses. The hormones may come from Mom or from the fetus's own gonads—tiny testes or ovaries—and, like seasonings in the amniotic fluid, they bathe the fetus's neuronal circuits. Hormones affect the organization of the mind, building up some regions and weakening others. They also control the development of the testes and ovaries. And, as it turns out, they also affect the relative growth of the fingers. The digit ratio can be seen as a sort of high-water mark left behind by the hormonal tide.

I am delighted to see that our daughter has approximately even ring and index fingers. "Aha, it looks like she's going to be verbal like us," I announce. She has a digit ratio that is typical for many girls and women.

People whose ring fingers are *longer* than their index fingers—a marker of high prenatal testosterone exposure—tend to have a male-typical pattern of brain organization. This is associated with athleticism, assertiveness, good spatiovisual ability, high numerical but low verbal aptitudes, and musical talent. Long-ring-finger types are more likely to have more efficient hearts and stronger musculature. In study after study, they perform better in soccer, endurance running, skiing, dancing, and so on. Towel-snapping jocks, they might also be driven to push themselves physically and be more competitive. Women with longer ring fingers are more likely to be bisexual. They may be more likely to be fresh air fiends and lash out when provoked.

Interestingly, palmists have long associated the ring finger with the heart, following the ancient Roman belief that this, the fourth digit of the hand, contains the vena amoris, the vein of love connecting the hand to the heart—explaining why the wedding ring is worn on this finger. Males with well-endowed ring fingers are more likely to have a raging libido, a higher-than-average sperm count, and more powerful-looking facial features, such as prominent eye ridges and rugged jaws. Casanova was known to brag about the length and girth of his ring finger.

Generally, males have ring fingers about 4 percent longer than their index fingers, and females often have index and ring fingers of about the same length or with the index finger slightly longer. A woman who has a fraternal male twin is more likely than her peers to have this finger length proportion and the masculinized features and behaviors that go with it—because she has been exposed in utero to her brother's hormones. While butch lesbians often have longer ring fingers, digit ratio is not a good predictor of male homosexuality.

(An interesting aside: A better predictor of a man's sexuality is the number of older brothers he has. If you're carrying a son, you may mount an immune response against male-associated proteins acquired from being pregnant previously with a male. That immune flare-up may affect the development of the fetus's brain and body. The chance of a male's growing up to be gay increases by 33 percent with each biological older brother and accounts for 15 to 30 percent of homosexuality in men.)

A longer index than ring finger is associated with verbal fluency, the ability that makes a person good at naming fifty types of fruit and reciting poetry. The longer the index finger is in proportion to the ring finger, the less prenatal testosterone and (perhaps) the higher the levels of estrogen. Among women, a longer index than ring finger is also associated with high fertility and a tinier waist, a plumper lower lip, a narrower nose, and a more delicate jaw line. Longer-index-finger types are more likely than longer-ring-finger types to get upset seeing other people cry. Our hearts race faster when we get emotional. We have a soft spot for sufferers and score higher on tests of moral judgment. We're also more tolerant of wailing infants. Apparently.

But I've got to wonder, is all this a modern form of palm reading? On one hand, the associations between fingers and various traits are based on serious research. On the other, as far as telling me anything truly predictive about my daughter, myself, or anyone else, I need to get a grip. Will my baby grow up to be a star trader, a lacrosse player? A musician, a lesbian? It's not the size of her fingers; it's what she does with them. And some answers best remain in the hands of fate.

7

EVE'S LEGACY, NIPPLE POWER, AND THE GOLDEN HOUR

Some Science of the Maternity Ward

Ten days past my due date, the baby I have spoken of so lovingly has become "the fetus."

The downgrade to improper noun means I'm peeved. The fetus is giving me backache, heartburn, edema, night sweats, hemorrhoids, mild incontinence, and stretch marks. People do a double take when they see me. "Why are *you* outside?" a stranger on the street implores. I'll never forget her shocked face. A homeless man turns to another and says, "Daaaaaamn, beetch gonna have dat baby on da street!" Nights are spent in bed, sweating heavily and obsessing (about cobwebs in corners, thank-you notes, the effects of air pollution on fetuses, stillborns—that kind of thing). I get up at five to sit on the couch, where I lay my hands on my immense belly and watch the sun rise over an overripe city. A yellowish crud (colostrum, or "first milk") collects in the crannies of my nipples. I get tiny contractions, known as Braxton-Hicks, but they're teases. I fart extravagantly, as if the fetus is clenching my stomach and intestines. She's giving me a raspberry. She's mocking me.

Even so, I am sappily sentimental about the fetus. She is already everything to me. I well up at the most manipulatively maudlin triggers. An ad for a bassinet that converts into a bed for a baby girl that grows up. My mother telling me my birth story, and how, when I will look into my own little newborn's eyes and feel her tight little grip, I'll know unconditional love. I'm dimly aware I've lost it after reading an article about monkeys. Sometimes, when an infant monkey passes, its mourning mother may carry the body on her back for weeks, grooming it occasionally, unwilling to let go. I cry and cry. It's just the hormones, the scientist in me reminds myself. Progesterone and oxytocin are schmaltzy bonding hormones, and I'm saturated with the stuff. This and sleep deprivation make me easy prey.

Only the baby's birth and all the practicalities that follow could stop this nonsense.

What triggers labor remains one of the great mysteries in life. "Birth is a result of complex, partially defined, events that are tightly regulated by a variety of mechanisms and mediators of endocrine, nervous, and immune systems," one study blithely asserts and then adds: "Unfortunately, none of them is completely elucidated."

Most scientists would agree on this much: the birth process begins with a covenant among the various hormone-producing players: the fetal hypothalamus, the pituitary glands of mother and fetus, and the placenta. Up until the last days of a pregnancy, the uterus is a mostly peaceful, nurturing refuge. The hormone progesterone, produced by the placenta, has maintained this Eden. The weather changes soon after the placenta sends out a stress signal, corticotrophin-release hormone (CRH). This signal stimulates the fetus to start producing steroids. Once these fetal steroids are in circulation, the placenta converts them into estriol. Estriol suppresses progesterone that has until now preserved paradise in the uterus. All hell breaks loose. After this, the pituitary gland produces oxytocin, which gives rise to fierce contractions. Prostaglandins, hormones also produced by the placenta, open the floodgates by helping the cervix soften and dilate.

Then the fetus is sent on her way. And it's not a gentle good-bye.

————————

Finally, at nearly forty-two weeks, I am zapped with lightning-like cramps. They intensify and become more frequent over the course of the day—from once an hour, to every half-hour, to every ten minutes or less by evening. All night the storm rolls in. I surf the Web for advice on how to ease labor pains. I unfurl a yoga mat and attempt cat-cow poses, peering out the glass door into the humid night. I moan. I meditate. I breathe. I hyperventilate. Sometime after dawn, a thunderbolt contraction strikes, sending me reeling and retching. I wake my husband.

A series of events follow: a voiding of the bowels and stomach followed by a stumbling exodus to the steaming street, morning pigeons everywhere, and into a taxi. Breathing deeply and deliberately, I am spirited to the hospital. I strip, shakily don a paper gown, and settle on a throne-like recliner in the center of a birthing room. I am the queen bee. Swarms of people come in to attend to some matter or another—to draw blood, collect urine, offer water, ask questions, insert an IV, and plug in the epidural.

I have great respect for women who give birth without medication, my mother among them, but I choose to blunt the pain. ("I've never had a spinal injection before," I confide to the young lean-jawed nurse who is setting up my epidural. "I haven't either," she says, wheeling a prep table over. "Yah, it'd freak me out too.") It turns out the sensation of the needle sliding between my vertebrae is like a mosquito bite compared to the pain of the contractions. Ten minutes later I relax.

I breathe. Beyond this, I cannot account for the long hours I spend as my husband dozes bedside and the light drifts across the room and fades. I am less stressed, more carefree, than I thought I'd be. I can feel the contractions despite the epidural, which, as it turns out, works on my right side only. The anesthesiologist offers to fix it, but I like it, it's perfect. I have feeling, but the edge is dulled. By early afternoon, my amniotic fluid is still intact and my cervix has not dilated much. My doctor breaks the water with what looks like a crochet needle. The nurse would like to induce with synthetic oxytocin, but I refuse. More time passes.

At dusk the contractions intensify. Peter wakes up from his nap and smiles leisurely. He takes in a meal. He calls his friends. He's feeling well.

By now the pain has returned and I know it's time to push. After so many hours of watching the sun slide across the room, of breathing and sipping water, we're suddenly on. Out of nowhere a mob materializes: the obstetrician, nurses, a medical student, and other phantoms in white coats. I ask for Bach's *Cello Suite*. Knees to nose, all is exposed, but this is no time for modesty. With each contraction I am to push to the count of ten. We begin. One-two-three-four . . . I hear the cello. Breathing in, breathing out.

Forty-five minutes of pushing and puffing pass, punctuated by long bouts of friendly foot shuffling as everyone in the room listens to the music and waits for my next contraction. After fifteen or so rounds, one of the nurses, I don't remember which, sees hair: "I see hair!" It's like seeing land when you've been on a ship for months. The whole crew cheers. The obstetrician reaches for the overhead klieg. Peter grabs his Nikon. I become strangely slaphappy. *"Lights, camera, action,"* I chirp doltishly. "Guess I'm responsible for the action."

So I push. And push. My knees are trembling as the baby's head pops out from behind my pubic bone. If there is a point of no return, it's now. The baby is stuck up there, in limbo, half in this world and half in another. My eyes are bulging in pain. "I know what they mean by ring of fire," I say gamely. Now dilated to a full ten centimeters, a circus lion could jump through my cervix. Prostaglandins and progesterone and oxytocin, oh my! One, two, three, and four . . .

I close my eyes and drift into a sort of daze. Bach is cracking the whip. The cellos urge on. The chords tighten, the chords loosen. Breathing in, breathing out.

PUSH!

"One more strong push and we'll have a head," the obstetrician says mildly. I let out a gargling laugh and open my eyes. I'm blinded by the light.

PUSH!

I close my eyes and focus on the image of a blazing ring. It's probably the afterimage of the overhead light seared into my retina, but it feels sacred. The ring turns into a vacuum, and the vacuum fills my entire field of vision. Even the music is sucked through it. And now it's baby's turn.

PUSH! PUSH! PUSH!

Do I remember the pain? Yes, but only in a hazy and muffled way. The pain is irrelevant the moment the doctor puts the fetus on my chest, her umbilical cord dragging like a tail. Seven pounds, nine ounces of fetus. We look at each other. I am smitten and startled.

Of course, I can't call her the fetus any longer. She's a baby now, with my lips and my husband's flashing almond eyes.

Her name is Una Joy.

> The power and intensity of your contractions
> cannot be stronger than you because they
> are you.
>
> —Anonymous

IS THERE A PURPOSE TO PAINFUL BIRTH?

To punish Eve for eating the forbidden fruit of knowledge, a vindictive God not only banished her from paradise but made a decree. "With pain," He said, "you will give birth to children." What He meant: the price of brains is pains. A gorilla gives birth in thirty minutes. A woman is in active labor for an average of twelve hours.

Taken metaphorically, Eve's pain-for-brains trade-off happens to jibe nicely with a rather more scientific theory. The human skull, encasing a brain that is larger and superior in intelligence to those of our ancient ancestors, can barely fit through a woman's shockingly narrow pelvis. You might think our pelvises would have expanded too, over the course of evolution, but no. If the pelvis were any wider, we'd have a wide, lumbering, orangutan-like gait that wouldn't get us very far very fast. There's not much leeway between a pelvis big enough for birthing a big-brained baby and small enough for walking her when she cries. This is the downside of being upright.

And so, to be born, a baby must navigate the tightest and trickiest passage of his lifetime. The birth canal is wider from side to side at the top, near the uterus, and then, at the bottom near the exit, it's wider from front to back. It's not only a tight squeeze, it's also twisted in the middle. What does this mean? Pain, peril, and pointy-headedness. Like working her way into a long boot, as midwives describe it, the tiny contortionist must rotate enough so that her head is at the widest part at the bottom and the shoulders are at the widest part at the top. A baby can get stuck in the pelvic inlet if his head is too big, in the wrong position, or doesn't corkscrew correctly.

Hormones such as relaxin loosen the ligaments and help widen the pubic symphysis, the space where the baby exits the pelvis. This Open Sesame event is so dramatic that it scars us internally. Even weeks after giving birth, I feel as if my pelvis has been blown open by dynamite. Experts can tell how many children a woman has had by counting these scars on her skeleton, like notches on a bedpost.

The "brainy biped—puny pelvis" theory has been the most popular explanation for why childbirth is so painful, but it's not the only one. Another is that the labor pain has an evolutionary purpose: it makes birth more emotionally significant, which increases the chance that the newborn will survive. Emotionally charged events bond people together. The mother's community—family, friends, other women, the baby's father—rallies to help her through the process. Bonding fosters trust and stability within the community. Babies born in this context have a higher chance of surviving infancy. The baby survives, and the species thrives. Ergo, human evolution favors painful births.

This theory may sound farfetched, but one part of it isn't controversial: women almost always require assistance during childbirth. Rare is the lady who can pop one out without help (in fewer than 10 percent of nearly three hundred cultures surveyed does unassisted birth happen, and the woman is almost always a veteran mother). The reason for such dependence is another side effect of our twisted pelvic anatomy. While nonhuman primates are born face-up, most humans come into this world face-down. We're born turning our backs to our mothers. In contrast, a laboring gorilla could bend over, peer between her legs, and make eye contact with her baby. She could help him out, and then he'd clamber up her body toward her nipples. An unassisted human mother, attempting to reach down and do the same, might accidentally snap her newborn's backbone or crush his windpipe. Relatively speaking, the human baby is a weak-necked limp-limbed deadweight.

It's quite something, the pain related to our awkward reproductive anatomy. Only half of us Americans who intend on having a so-called natural, drug-free birth actually go through with it. Even with the right side of my uterus numbed from a half-effective epidural, some of the contractions were eye-popping.

Perhaps we Westerners who numb our nerves at the slightest shade of a headache are wimpier than our foremothers—women who, even in their most excruciating moments, had only a rag and bone to bite on. Epidurals are uncommon in Asia, the Middle East, and Africa, where women are expected to tough it out, yet about 70 percent of American women have them.

Is pain just a state of mind? Expect rip-roaring agony, and you'll surely feel it. Fear makes us more sensitive to pain. Have some faith in your body, natural birthers insist. Our pain threshold goes up in the third trimester (we hardly feel the cold), and labor at term activates inhibitory neurons in the brain. Many claim childbirth is ecstatic pain, like running a marathon, and the soreness is chased away by a natural endorphin high. Even so, some of us have lower pain thresholds. If you're a redhead, you may be less tolerant of pain due to a variant of a hormone receptor. If you're older than the average pregnant woman (over thirty), you may have less intense pain.

Then there's the question that aches for an answer: Was childbirth always this painful? In early cultures, childbirth was probably easier because people were in better shape and accustomed to physical labor and perhaps less fearful of pain. An oddball theory introduced by the nutritionist and scholar Weston Price, based on anthropological evidence, claims that since the agricultural revolution human pelvic inlets became narrower and shallower and our skeletons less robust. Nutrition, or lack of it, claims Price, resulted in these anatomical changes as our diets shifted from protein-rich fats to carbohydrate-heavy grains. If the hunter-gatherer world of the past was something of a Garden of Eden, the theory goes, then childbirth became much more painful when we left it behind to sow our own seeds.

And yet labor pain can be managed. We tap into our tree of knowledge: meditation, midwives, supportive friends and relatives, Lamaze breathing, yoga, music therapy, warm baths, acupuncture, massage, modern painkillers—or whatever other means of relief our big brains have conceived. We do what we can. It's Eve's burden, borne by every childbearing woman.

The Science of Squat

Our ancient foremothers had their way of dealing with the brainy biped–puny pelvis dilemma. They almost certainly gave birth upright, in a squatting position. "If you lie down," the Native American saying goes, "the baby will never come out."

This much is proven: anatomically speaking, a seated, splay-kneed pose is the easiest way to squeeze out a fetus. Assuming the baby is head down, squatting puts most of the pressure on her occiput, the cranial plate at the back of the skull. This decreases the risk of injury, such as shoulder dislocation, because the biggest, sturdiest part paves the way. Squatting also expands our skinny pelvises by 20 to 30 percent more in diameter than any other position, and it increases pressure on the abdomen, making it easier to push. We're harnessing gravity too. Compared to the usual recumbent approach to childbirth, upright labor is 25 percent faster, decreases the risk of having to stretch and extend the vagina with an episiotomy, minimizes perineal tears and blood loss, reduces the need to medically induce, and causes less severe pain.

The only catch, as anthropologists point out, is that women in Western societies sit, not squat. This is a modern problem. For most of human history, we'd crouch while we hung out, cooking, gathering, gossiping, and so on. Squatting would come naturally. Now most of us deliver on our backs, the position that works better for obstetricians and women who choose epidurals or other painkillers that make them weak in the knees. But for those of us considering natural labor, squatting is clearly the best option. Reborn by modern science, it brings us closer to our roots.

DO WHITE MAMAS HAVE LONGER PREGNANCIES?

A full-term pregnancy officially begins at thirty-seven weeks. This is three weeks before our estimated due dates, but if we happen to go into labor sometime during week 37, we still get full credit for a full-term pregnancy. By thirty-seven weeks, I was hoping I'd go into labor—I had overcome much of my fear of childbirth—but my body and baby were holding back. I was disappointed. Then weeks 38, 39, 40, and 41 passed and still no baby.

My husband and I are both procrastinators, so we saw the delay as proof of parentage. It must be in the genes, we joked.

A doula at the yoga center waved away any crazy theories I had about the baby's temperament. "You're late because you're white," she said. I thought she was kidding. Then, talking with her, I was made to understand that she wasn't. A little secret of those in birthing professions is that race influences the length of gestation. White babies are the pokiest, and Asian and black babies are the speediest.

The difference between a slow and speedy pregnancy isn't much: a week on average. The delay doesn't matter to anyone except the desperately pregnant woman and perhaps her doctor or doula. The race effect has been long known anecdotally, as a sort of old wives' tale, but it's been convincingly proven in a recent analysis of over 122,000 pregnant Londoners. Crunching a decade of data, researchers found that while the average pregnancy is officially forty weeks, or 280 days from the date of the last menstrual period, this is true of white women only. Black and Asian (including Indian and Pakistani) women, it turned out, enjoy speedier pregnancies that average about thirty-nine weeks. They are also 33 percent more likely to deliver preterm than white women (other races were not included). This was found regardless of body mass, socioeconomic class, cigarette smoking, and other factors.

The theory behind the racial difference is fascinating, White babies take a bit longer to mature. This means that a black or Asian fetus at, say, thirty-four weeks gestation is likely to be more developed than a white baby of the same age, and therefore have a survival advantage. All else being equal, a white preemie is also more likely to stay in the neonatal intensive care unit longer than a black or Asian preemie. After birth, black babies have also been found to develop gross motor skills sooner than white babies. Nonwhite babies may have a head start and a leg up, literally.

One sign of maturity that doctors look for is meconium in the amniotic fluid. Meconium is a black-green sludge that suggests the baby's digestive system is going online. (Meaning "opium-like" for its color, meconium contains everything the baby swallowed as a fetus, usually its cells, hair, bile, and mucus. In the beginning, babies literally eat themselves. Later, through breast milk, they eat their mothers.) The stuff, fearsome to look at but sterile and odorless, is what we will soon find in our newborn's diaper. In postterm babies it often colors the amniotic fluid. In black and Asian moms, there's a somewhat higher chance that meconium will be present in their water by their due date, and even well before their due date.

So why would black and Asian babies grow a bit faster than white babies? That's the mystery, and the British researchers have a theory: pelvic variation. Pelvises vary (slightly) by race. On average, black women have slightly narrower pelvic inlets than white women. A narrower pelvis offers more stability when running, among other advantages. This wouldn't be a problem if human babies weren't born with such big brains. The longer the gestation, the bigger the head. A noggin that is too big may obstruct labor or cause too much pain in childbirth. To resolve the conflict between mom and fetus, evolution may have made babies of moms with slightly smaller pelvic inlets mature slightly faster, so they may be born before they get too big.

What does this mean for mixed-race couples? Some studies have found that the mother's race matters more than the father's when it comes to the length of a pregnancy. A baby born of a white mom and black dad is more likely to go the full forty weeks than a baby whose mom is black and dad is white. Whatever genes mediate fetal maturity may be more influential when they come from her—as if nature custom-fits the fetus for the pelvis. Age, previous pregnancies, medical complications, cigarette smoking, stress, and social factors may also play a role in short and preterm pregnancies. But race shouldn't be overlooked in how quickly a pregnancy will reach the finish line.

Is Daddy Delaying Us?

The placenta that protects and feeds the baby is not exactly on our side. We have shockingly little control over the organ. It works for us, but it also works against us. It's manipulative, and it takes what it wants without asking. Should it surprise us that it's run by male genes? Maternal genes are in the placenta too, but many (not all) of them are silenced (imprinted). Dad's genes dominate.

The reason that placentas are paternally biased is that evolution doesn't trust maternal genes to run the whole show. Let's imagine our tired and drained bodies decide to end a pregnancy early. Good luck persuading Papa Placenta, which plays a role in deciding when labor starts. Dad's genes want to take as much as they can and for as long as possible. (After all, our next pregnancy might be with another man.) Unfortunately, he might have some control over the placental clock.

The father's genes may increase the risk of a long (forty-two-week-plus) pregnancy by 30 percent, according to one major Danish study that looked at twenty-two thousand women who had drawn-out pregnancies for their first child. Other studies suggest that fathers who contribute genes that make a baby slightly faster growing (black men, perhaps) may influence a shorter gestation. All of this implies that the father's genes play a role in delayed labor even though the mom's nonplacental genes are more influential overall. And some dads' genes, it seems, dawdle more than others.

> Those wretched babies don't come out
> until they're ready.
>
> —Queen Elizabeth II

CAN WE INDUCE OURSELVES?

Everyone who saw me bumbling along at week 42 openly pitied me. Most offered a do-it-yourself induction strategy: to eat/drink/chant/prick/visualize/believe something to speed up the baby's birth. I was big, so I was game. Home induction seems so much gentler than medical induction. It would be an encouragement more than an eviction.

I dismissed "techniques" that seemed crassly commercial—eggplant parmesan and a cream cheese spread touted by restaurants as induction specials . Herbal remedies I also disqualified—not because I didn't think they'd work but because they require expert oversight. It didn't take long to list the big six: walking, spicy food, castor oil, sex, acupuncture, and nipple stimulation.

Walking is the most commonly recommended strategy. Two out of every three of us, according to one study, believe that walking brings on labor. A daily constitutional strikes me as commonsensical, and benefits include positioning the baby in the pelvis, stimulating muscles, and alleviating anxiety. But, medically speaking, will waddling around really trigger labor? Probably not. Walking didn't induce labor for me, and I took a mile-long ramble every day of the last six weeks in my extra-wide sneakers and stretch shorts. In theory, dehydration (due to exercise) may decrease blood volume, which in turn increases the level of contraction-causing oxytocin. But this doesn't sound wise, so I abstained. Moving on.

Spicy food and castor oil theoretically work by irritating the digestive tract. The idea is that gastrointestinal spasms trigger uterine contractions that trigger labor. Castor oil, made from a flowering plant, *Ricinus communis,* may also work by inspiring the body to produce prostaglandins, hormones that rouse the uterus like a motorboat making waves in a quiet lake. While the effectiveness of spicy food has not been scientifically studied—and likely varies with personal background and tolerance (why would it work for cast-iron-stomached curry-eaters?)—there is limited support that castor oil does. One study found that among women forty to forty-two weeks pregnant, those who took two fluid ounces were significantly more likely to go into labor within twenty-four hours than a control group (75 percent versus 58 percent). But the downside to castor oil concerns the backside. The dirty secret behind the stuff is that it causes a complete voiding of the bowels, if not the uterus, and women who take it deliver more than desired in the birthing room.

The castor oil tactic should not be combined with the next one: sex. But the thinking behind it is not much different: uterine contractions, from orgasm in this case, may help trigger labor. Prostaglandins in semen would also help the cervix "ripen" and widen. While it's a valid theory, there's no scientific proof that induction sex is anything but a labor of love. Several studies have found no difference in spontaneous labor among women who had sex at full term (most women had an orgasm at least once at that late stage of pregnancy) than in those who abstained. Interestingly, sex was associated with a decreased risk of preterm labor in one study of women between twenty-nine and thirty-six weeks. For whatever it's worth, my late-pregnancy orgasms did not seem to get me any closer to labor. They'd just leave me with a dull ache in the abdomen. One lesson is clear: those who get it on don't necessarily get it out.

Another penetrative act, acupuncture, doesn't fare much better when put to the test. Slipping thin needles deeply into the skin at particular nerve points is thought to stimulate the nervous system directly (or indirectly, by increasing oxytocin production), which leads to uterine contractions. Although I didn't try to induce labor with acupuncture, I did use it when trying to conceive. Okay, I got pregnant and the baby turned, but I do not know whether to credit the needle because I tried many other techniques at the same time. Similarly, there's no strong evidence that acupuncture really works, or not, to trigger childbirth. Many studies have found that needled women take the same amount of time as the needleless to go into labor, so the treatment seems needless. But acupuncture's effectiveness is hard to pinpoint given how much variation there is in the duration of treatment, number of points used, depth of needling, and points selected.

This left me with one more home remedy: nipple stimulation. More pleasurable than needles and castor oil, this maneuver involves stimulating—that is, sucking, pumping, and pulping—the nipples and breasts. Nipple stimulation triggers oxytocin production, which makes the uterus contract and expel. Several studies have found that nipple stimulation makes the uterus contract stronger and the cervix ripen faster than it would otherwise. But rubbing yourself raw on the order of three hours daily for three days is labor in its own right. While the deed can be done with a breast pump or hands, it's more pleasurable to recruit a partner. Incidentally, oxytocin is a hormone that makes people more open and trusting—including falling in love, which may be helpful at this time. And the steady, nagging tug prepares you for breast feeding. Sooner rather than later, hopefully.

I spent two hours sheepishly trying manual nipple stimulation and went into labor a day later. At that point, I was nearly forty-two weeks along. Was it the pumping or the timing?

This much is clear: self-induction is unlikely to work on a woman whose baby and body aren't already close to being ready for labor. Yet the willingness to try these methods—that alone is a sign of readiness.

> However long the night, the dawn will break.
>
> —African proverb

WHY DOES LABOR (OFTEN) STRIKE AT NIGHT?

My labor started at dusk. Cramping and sweating, I never before felt so much in touch with nature. I think it had something to do with the hour and my solitude and being in pain that I could not turn off or transcend. I've heard many stories of women able to meditate or hypnotize themselves through the contractions, but I couldn't invoke those higher brain functions, that higher self, not for more than a minute. So I howled and hissed and growled and grunted all through the night. By dawn, I was on the floor, bowed over my big belly and breathing hard. I was an animal. I don't mean this in a bad way.

This is how labor begins for many of us—contractions that begin after sunset that intensify through the night. Although the timing of actual onset is as impossible to pinpoint as the day the earth began—labor is a buildup, not a distinct moment—most studies agree that meaningful contractions more often begin somewhere between midnight and 4:00 a.m. This doesn't mean that all women start labor then, but it happens to enough of us to be significant. I wonder: deep within our human nature, is there a reason that women often go into labor alone in the dead of night?

The first clue involves our circadian rhythm—the twenty-four-hour cycle of night and day that affects our bodies and behavior. Hormones are sensitive to these rhythms, and among those that peak at night are estriol and oxytocin. Estriol sets the scene for labor to happen. As estriol rises, progesterone sets. Progesterone is the custodian of a calm, contractionless uterus, and when it dips beneath the horizon, the chaos begins. Meanwhile, like singles at last call, oxytocin receptors in the uterus are more concentrated and receptive in the wee hours of the morning. Oxytocin makes uteruses clench and cervixes dilate. (In synthetic form, oxytocin is called pitocin, which is used to induce labor.)

Why are estriol and oxytocin such ladies of the night? Maybe there's no explanation. But it's more interesting to look for a deeper meaning, which is exactly what evolutionary psychologists have done.

One observation they've made is that other mammals that are awake during the day—chimps, dogs, and cats included—also tend to start labor at night. Nocturnal mammals tend to start labor during the day. Why would we all start a task so painful and risky when we'd otherwise be resting? Because that's when we're safest, calmest, and least distracted. Our foremothers may have found night labor advantageous because other members of their tribe were home at those hours—no one gathers food at night—and available to help and protect them.

Times have changed, but we still need safety and calm. According to the prenatal ecologist Michel Odent, a modern laboring woman must be able to cut herself off from the world, to hiss and fuss and "be on another planet." This includes shelter from the sun and any other visual stimulation. In the silent dark of night, our rational self may release its grip on us. We can tap into the primitive part of the brain that releases birthing hormones. And we need privacy, which night's shroud offers, because being observed make us anxious and vulnerable. People in traditional societies go into seclusion when they labor. Monkeys peel off to find private groves, and cats and foxes find dens where predators can't see them.

This brings us back to the hormone oxytocin. When we're threatened or overstimulated, our bodies are flooded with oxytocin's adversary, adrenaline. "Fight-or-flight" adrenaline and labor-inducing oxytocin are like oil and water; they don't mix, and when one is up, the other is down. The more stimulated we are, the more adrenaline we produce. The more adrenaline, the less oxytocin. The less oxytocin, the fewer and milder the contractions and the more prolonged the labor. Get distracted or scared or angry in the midst of labor, and your mighty contractions whimper away. Incidentally, the same goes for orgasms.

This may explain what happened to me after a night of mouth-drying, gut-compressing early labor pains. Convinced I was about to give birth imminently, my husband and I rushed out to the street, where we tried, with increasing desperation, to catch a taxi to the hospital. Ten minutes later we found one, but not before I panicked, thinking the baby would be born right there, at the corner of Twenty-Third Street and Eighth Avenue in Manhattan.

By the time I stumbled into a cab, a strange thing happened: the contractions slowed down. Okay, I still could not speak during them, but they were not nearly as intense or as frequent as those I had in my living room. In retrospect, I believe some animal instinct kicked in, suspending labor until I felt safe, calm, and undistracted. Had my labor not been interrupted, I might have delivered in the morning, as many other mammals do.

My contractions were so quiet after that stress that the nurses pushed artificial oxytocin, which I refused with a tired wave of the hand. Hours passed. The sun drifted across my room and faded. I huddled under a blanket and breathed. As the lazy afternoon settled into dusk, my contractions picked up again, and I started to see the light.

> If pregnancy were a book they would cut the last two chapters.
>
> —Nora Ephron

WHAT HAPPENS IN THE GOLDEN HOUR AFTER BIRTH?

My husband takes four frenzied photos the moment our daughter leaves the birth canal. *Shot one:* my hair, those long tresses that I have not washed for days, hangs in matted clumps. I am wearing a snowflake-patterned institutional gown that gapes in the back and am tethered to IV lines. I smile unsteadily. *Shot two*: a cream-and-crimson blur. *Shot three*: the blur turns out to be the baby being placed on my chest. My chest is covered by a towel and padding. The cracked corner of my mouth is in the frame, and I appear to be grinning. *Shot four:* the foolish grin remains on my face but the baby is gone.

Total time lapse between photos one and four: seven minutes.

Where did the baby go? First to get washed, and then to be weighed and footprinted as the obstetrician fishes out the placenta. Eventually the baby is handed back to me, swaddled and asleep, for photos with the grandparents and our wheelchair ride to the postpartum ward. But there we part again, me to my room to recover, and she to the nursery for a more thorough cleaning and appraisal. We'll reunite hours later. My baby's first full hours on the planet are spent with strangers.

If there is an hour of my life that I'd take back to do all over again, it's this, the golden hour, the first sixty minutes after giving birth. Dozens of well-designed studies have since convinced me that I should have done what every other mammal has ever done: press the newborn baby, bare and even bloody, against my naked chest and keep her that way for at least an hour. The pose is sometimes called "kangaroo care," after the way that the mother marsupial carries her joey in a pouch. The baby is belly down, her torso between her mother's breasts, her head tucked under the mother's neck. Skin-to-skin, heart-to-heart.

In rat studies, newborn pups who are separated from their mothers let out pathetic little cries of distress as if they are being tortured. Their stress hormone levels skyrocket. Over a long period of time, they become sickly and brain damaged. Bleeding-heart researchers rage against these animal experiments because they're cruel—an irony, because in many hospitals newborn humans are isolated from their moms. Especially after a C-section, many women don't see their baby for hours, or even a day, after giving birth.

This is what's ideally supposed to happen in the first hour of life. Your newborn will look at your face and get a hazy sense of its outline. If you stick out your tongue, she may laboriously stick out hers in kind—this is the beginning of a relationship. She'll snuffle at your chest and squirm toward your nipples. The memory of your odor sears her memory circuits.

Minutes after birth, babies have a supersensitive sense of smell, and whatever odors they encounter, even unfamiliar ones, are imprinted in their memory. When experimenters exposed babies to cherry and mango oils the hour they were born, the babies remembered these smells days later and preferred them. This window isn't open for very long. When the experimenters exposed babies to cherry and mango odors twelve hours after birth, they showed no memory of or preference for them later.

Why does the window slam shut so quickly? The stress hormone cortisol that a newborn produces at birth activates her central nervous system, including the smell and memory regions of her brain. As cortisol declines rapidly in the first hour after birth, so do the baby's olfactory powers. This is why newborns may grow attached to the first people they smell. They're programmed to recognize and prefer their mothers and other caregivers. Like cement drying, there's only a certain amount of time to leave a first impression.

Of course, impressions go both ways. As the newborn roots around your chest, she sets off a hormonal trip wire. Two of those hormones are oxytocin and prolactin, which make you produce milk. If you spend the golden hour in an intimate embrace with your newborn, you're eight times more likely to breast-feed spontaneously than a mother who doesn't. Your milk supply will likely be established sooner, and a virtuous cycle of supply and demand begins, which may be why babies who had skin-to-skin contact in the first hour of their lives are more likely to be breast-feeding four months later. These babies may be better latchers and regain their birth weight quickly (babies usually lose weight until Mom's milk comes in). A baby who spends her first hour on her mom's (or even dad's) chest doesn't usually cry for more than a minute, unlike a separated and swaddled baby.

But crying may have a purpose in the golden hour too. In a baby's first tears, a chemical cue may be transferred to his mother. A fascinating new study by Israeli researchers found that tears contain hundreds of chemicals (incuding prolactin) and that some of them may act as chemical "mind-control" signals (or pheromones) on those who come in contact. In their experiment, male volunteers who sniffed tear-soaked pads harvested from emotionally distraught women had lower testosterone levels and reduced sexual aggression. It's not hard to imagine that a newborn's tears can manipulate his mother and father or other caregivers into producing higher levels of prolactin and oxytocin, the "cuddle hormones." The higher a person's levels of these hormones, the stronger the bonding instinct.

Yes, from the suckling baby's perspective it's helpful for us to start producing oxytocin, which makes us more nurturing. It works like a charm. Within minutes of holding my daughter, tears were streaming down my face. The nurse told me to get some sleep while the baby was being evaluated, but how could I do that? We evolved to spend this time bonding.

A blindfolded mother who has spent the golden hour with her newborn can identify the baby by touch alone (fathers take six to seven hours of contact). If our foremothers took any longer to bond with their babies, none of us would be here today. An indifferent mom would probably lick her wounds and meander off, leaving her young to die of cold or hunger, or become a predator's lunch.

The warmth you radiate in this hour after birth is more than emotional. The oxytocin rush raises the temperature of your chest, and in placing the baby there, you warm him. Imagine early ancestors in their caves and treetops, clutching their shivery newborns to their bosoms to keep them alive. Studies have found that babies who spend their first hour on their mother's chest have a core temperature at least one degree warmer than those in an incubator. As I write this, there is news of an Australian mother whose premature newborn was deemed dead at birth. His body was handed to her to say farewell. She warmed the lifeless baby on her naked breast and he started to squirm.

Toasty and relaxed, sniffing and suckling, skin-to-skin babies make more flexion movements several hours after birth. This means they bend their legs at the knees, the thighs resting on or near their stomachs. Then they close their hands and fall asleep. Flexing this way is a sign of a soothed central nervous system in a newborn, a position that yogis admiringly call "child's pose." Infants sent to the newborn nursery, in contrast, tend to extend their limbs akimbo like doing aerobics on acid. Advocates of "skin to skin" ask would-be parents: Would you rather your baby enter the world in peace and calm, or kicking and screaming? Skin-to-skin babies drift into deeper dreamlike states. They receive a warmer welcome to the world.

I sometimes wonder about this lost golden hour. Have I sacrificed any advantage in the long run by not having enough intimate contact during those first sixty minutes? For preemies receiving kangaroo care, the answer is clear because skin-to-skin is proven to help save their lives. For the healthy baby born at term, it's an open question because so few studies have explored the benefits beyond a few months after birth. One is a Russian experiment that assigned new mothers to either skin-to-skin or swaddle-and-separate groups within an hour of delivery. The researchers found that mothers and their infants in the skin-to-skin group had stronger, more responsive relationships one year later. How can we explain this? Maybe early intimacy leads to more breast feeding, which eventually leads to greater overall intimacy. Or early skin contact might inspire more of the same, and the benefits build over time.

Obviously, whatever happens in the first hour of life is added to or subtracted from as time ticks on. I take this as a lesson. Now when my baby is fussy, I unsnap and pull up her onesie, lay her on my naked belly, skin to skin, and feel her anxiety dissolve.

Our first golden hour together has passed, and I can't change that. Fortunately, there will be many more.

Why Is Baby Born Blue-Eyed?

When we gaze into our newborn's eyes, we often fall into a sea of blue. But the blue may be something of an illusion. It's probably not permanent. Many babies are born blue-eyed, and yet only a small percentage of them stay that way. The reason is that levels of melanin, the protein pigment in the iris of the eye, are low at birth, especially

In babies with white ancestry. A baby without much melanin has blue or dark gray eyes. A high-melanin newborn is black- or brown-eyed. Babies produce increasing amounts of melanin in their first year, and by their first birthday, their eye color has generally stabilized. Genes that determine the amount and type of melanin determine the baby's final eye color.

All permanently blue-eyed people on the planet can be traced back to an ancestor who lived six to ten thousand years ago in the Near East or the northeast part of the Black Sea region during the migration associated with the agricultural revolution. That person had a genetic mutation that limits the production of melanin in the iris. Before that, all humans probably had brown eyes. That babies of Caucasian descent are so often born blue-eyed has inspired an evolutionary theory to explain its advantage: fathers cannot disprove paternity at birth based on eye color, potentially reducing the risk of infanticide by angry cuckolds seeing red.

> Did you know babies are nauseated by the smell of a clean shirt?
> —**Jeff Foxworthy, comedian**

ARE WE STRANGERS TO OUR NEWBORNS?

Before the big gloved hands picked her up and spirited her away to be footprinted, before my husband held her, before the placenta was fished out of my body, and before they wiped me down and wheeled me away, I wanted to accomplish one thing. I desired to know if the baby recognized me.

I needed a sign.

Do those wet blue eyes see me? She's looking at me, but of course she wouldn't know my face. We lived together for about forty-two weeks, but this is our first tête-à-tête. Within two days after birth, infants might know the broad outline of their mother's face and hair and show preference for these features over those of strangers. In six to seven weeks, she should be able to pick my mug out of a lineup of head-scarved women. But right now, minutes after birth, she's legally blind.

Looking straight at me, she cries.

"Oh, sweetie," I whisper, stroking her tiny red face. "I love you." Could she recognize my voice? She should, according to several studies. In one, researchers recruited women and asked them to read *To Think I Saw It on Mulberry Street* by Dr. Seuss into a recorder. Within a day after the women gave birth, the researchers played those recordings to the newborns to see if they could distinguish their mom's voice from a stranger's. The babies, reclining in their bassinettes and wearing headphones, looked more oblivious than they were: they knew their mother's voice and showed it by sucking harder on the nipple of their pacifiers. Listening to Mom, they'd work the nipple harder than they would when hearing any other voice. (Someday they'll go through equal effort not to listen to their mothers.)

I look at my new baby. "I love you," I persist. She parts her lips and weakly cries some more. If she recognizes my voice, she doesn't deign to show it.

But she does rub her pug nose against the institutional cotton of my gown and snuffles. Do I imagine a flash of recognition across her miniature features? Her little face really does seem to brighten, and then it dawns on me.

She should know my smell.

My chest smells a bit like the only medium she's ever known: amniotic fluid. The chemical makeup of the broth that she's swallowed and marinated in for so long has some chemicals in common with my breast milk, sweat, and other secretions.

We all have a signature scent. It's made up of odors from our diets and chemical makeup. We also secrete volatile chemicals in our armpits and nipples, which may help the baby identify us (see page 13). Our smell is an anchor for our newborn babies; it makes them feel secure. My smell must seem somewhat familiar already to my baby, but the longer she rests and roots on my chest, the more she'll love and crave it, which is why we should never wash before holding our babies the first time. Given the chance to nurse, she'll learn to recognize me even more. This is bonding between infant and mom, where no substitute will ever do. Weeks later, another lactating mother will clasp my daughter to her shirt, and my baby will look half-aroused, half-confused. Infants prefer their own mother's milk.

Soon I'll know her smell too. All of us new moms in the maternity ward are addicts. We nuzzle our newborns—their necks, their chests, even their tushes. In one experiment, 90 percent of mothers who were exposed to their newborns for less than an hour identified their own baby's undershirt from others by smell alone. Mothers more accurately identify and prefer the smell of their biological children to their stepchildren.

It's during the sniffing and snorting and snuffling that something important happens. Not until later do I register its meaning.

The baby stopped crying.

An angelic peace passed over her face. Then the white latex hand swooped down and carried her off. But in that moment I had my sign. She knows me! My baby knows me from the inside out.

Why My Baby Smells Sweeter

Your own baby smells unlike any other. Your own baby also smells better than any other. This is the lowdown from an experiment that required new moms to sniff dirty diapers. One of the dirty diapers came from each woman's own baby, and the others came from strangers' babies. Even when the samples were deliberately mislabeled, women rated their own baby's fecal odor as better smelling (okay, less disgusting) than that of a stranger's baby. It's possible that mothers naturally adjust to their own baby's reek. It's also likely that we prefer our own baby's foul odors because they're a sign of biological relatedness.

"Sometimes when you pick up your child," writes Jodi Picoult in *Perfect Match*, "you can . . . smell the scent of your skin in the nape of his neck." The smell of sweat—and excrement—is influenced by the nuances of our diet and the bacteria that live in and on us. It's also colored by DNA, particularly a cluster of immune system genes called the major histocompatibility complex. Sniffing out people biologically related to us is useful from an evolutionary perspective. In the context of mating, women tend to prefer the bodily smells of strangers whose genes are mostly unlike her own. But when it comes to babies, we prefer body odors related to our own.

Preferring our own baby's smells, fair and foul, is part of bonding. In overcoming our natural disgust, we're better caregivers. We give preferential treatment to little ones who carry our genes, even stinky ones. So strong is this parental bias that we really think our own kid's s—t doesn't stink (well, as much).

DO WE REALLY FORGET THE PAIN?

"See you next year!" a maternity ward nurse jokes as twenty new mothers in hospital gowns hobble out of breast-feeding class. I'm not the only one who snorts in response. We resemble veterans of a successful but bloody and protracted campaign. "Advil or Tylenol with codeine?" the nurse offered each of us before the class began. I reach for the latter.

As time passes, the pain fades. The birth, it is said, wipes out all memory of steel-claw contractions, mashed fists, panicked retching, and the astonishing pressure in the pelvis as the bones spread. It all goes away when you hold your newborn, the story goes.

But do we really forget the pain? The question piqued the interest of Ulla Waldenstrom, a professor of midwifery at the Karolinska Institute in Sweden. Waldenstrom and her team approached nearly fourteen hundred women who had delivered their babies within the past two months. She asked them to rate their pain during childbirth on a scale of 1 to 7 and describe their feelings about their experience. Five years later, the researchers tracked down the women and asked them to rate their childbirth pain again.

Amnesia of sorts blessed about half the mothers, for they remembered childbirth as less painful than when they freshly delivered. This is the magic that we hope will happen, and the fact it works half the time is good news. Other studies arrived at similar results. These lucky moms with amnesia about their labor pains were more likely to have had positive birth experiences in the first place. As for the rest of the women, about one-third remembered their birth experience the same as they rated it five years ago, and about 15 percent remembered it as more painful. These latter women were less likely to have a second child.

What did Waldenstrom make of this? A lesson: we're more likely to forget the pain of childbirth if we had an overall positive experience. Our pain during and immediately after childbirth and the long-term memory of it involve different memory systems. The latter is colored by emotion. What do we think about when we think about childbirth? Do we remember being delighted or daunted? Were our husbands good cheerleaders, holding our hand when we wanted them to and giving us space when we didn't? Do we focus on the ecstasy of holding the newborn baby—or the humiliating poop on the delivery room table or the filmed-over blankness in the eyes of the nurse when we sought sympathy?

In a positive experience, the reward circuits in our brains are activated and release endorphins. They not only help us reduce or overlook pain in the moment, but also help us forget pain later. Moreover, we're likely to breed positive memories about the childbirth experience every time we think about it. These memories go forth and multiply. Nurse bad memories, and they too will grow.

The latest thinking is that memories aren't the permanent slabs of granite we like to think they are. They're more like dunes, whipped up and reshaped all the time. According to Joseph LeDoux, a neuroscientist specializing in the study of fear, an emotionally charged memory is only as good and as accurate as the last time we remember it. That is, the neurons in which memories reside are rebuilt every time the memory is recalled. New proteins are synthesized in the amygdala, the emotional command-and-control center of the brain, and in the hippocampus, where memories are consolidated. As memories are rebuilt (perhaps within a two-hour reconsolidation window after recalling a memory), new emotional information can be incorporated into an old memory. When you recall the memory, the content may be tinged by your most recent experience. It can alter in the telling and in the recalling. This is why memory isn't perfect.

Here's a thought. Our memory of childbirth may be especially susceptible to revision because it's an event we replay over and over again
in our minds and when talking to others. Every time we dish about the
delivery, our recollection of it may change a little. Others' experiences
and reactions may distort ours. Your memory may darken if you have
a partner who harps on how much you swore like a sailor when you
were in pain; it may brighten if he goes on about how brave you were
and how you gave him the best day of his life. If you had a C-section,
was it unexpected or planned? Women who have the latter remember
having far less pain. Your relationship with the baby may color your
recollection too. These new layers of meaning and context tint the fab-
ric of memory. The new blends with the old seamlessly. The memory of
physical pain depends on emotional context; if the emotional memory
changes, the pain memory may change, too.

That most of the women in Waldenstrom's study remembered
childbirth as less painful, or at least not more painful, five years afterward is a good omen. It suggests that the majority of us have positive
memories, or at least the pleasure outweighs the pain. Maybe the evolutionary explanation is true: childbirth amnesia helps our species
survive, because if pain memory were perfect, we wouldn't have more
than one child.

I admit it. As time passes and happiness increases, the pain seems
trivial. I'm one of the lucky ones. But will the maternity nurse see me
next year? Absolutely not. I'll know when I'm ready to have another
baby: if the thought of going through labor again is pure pleasure.

8

MOMMY BRAIN, MOOD MILK, AND THE WEIRD HALF-LIFE OF CELLS THE BABY LEFT BEHIND

A Postpartumology

I cannot sleep after giving birth. It is late, long past midnight, and the baby is being washed and evaluated in the nursery. My hospital room is half-dark, and I lay motionless on the bed, staring out the window at a view of the city's East River. It's summertime. A few private boats pass by, casting their glow on the glossy, black water. I feel shaky and light-headed. Everything has the texture of a dream.

Outside me, outside the room, I hear a squeak. The door opens, and in comes a full-figured nurse with a cart. She whispers hello and flips on the fluorescent light. The cart turns out to be a wheeled bassinette, and in the bassinette is a sleeping baby—my baby—in a pink-and-blue striped hat. It is 4:00 a.m.

"Time to try to breast-feed," the night nurse sings. She is professional but not unkind. She places the sleepy bundle in my arms and shows me how to position the baby's mouth. In a delicately awkward moment, she reaches into my hospital gown to jigger my nipple. "You want to squeeze your breast to make it bitable," she explains as she demonstrates on me. "It's easier for her to latch on if you make it more like a hamburger than a Sno-cone." I nod seriously. Dignity has gone out the door.

Looming larger is the problem of the baby falling asleep. The nurse and I whisperingly discuss carrot-and-stick methods of how to wake a newborn: stroke her cheek, undress her slowly, massage her, dangle her legs, douse her with cold water. We discuss foremilk and hindmilk (the former having more calories, the latter more fat) and the importance of emptying each breast before switching to the other side. The baby latches then unlatches and latches, slumbers and rouses. At last she settles on dreamily nibbling my nipple. By then the night nurse has turned off the light.

What is my newborn getting out of this? Not milk, not yet. Progesterone, the hormone that has kept the uterine lining tight and secure, like a military blockade around a castled city, repressed the milk-producing hormone prolactin. But progesterone's reign has reached an end. Soon prolactin will rise up and do its yeoman's work of extracting protein and sugar and a bit of fat from the blood and converting these nutrients to breast milk.

Milk delivery will take three to five days of suckling before baby's hard work pays off. She'll eventually drink about a quart daily. In the meantime, a tablespoon of colostrum is what's for breakfast. The stuff is a thick golden syrup, like popcorn butter. Or medicine. Colostrum packs a punch of antibodies and white blood cells to protect newborns against disease and growth factors to stimulate the gut. When I spent a season in India, I took a form of cow-made colostrum called Travelan to protect against the bacteria that cause diarrhea. The goo contains concentrated levels of protein and fat. It's also a laxative that flushes away meconium, the ferment found in a newborn's intestine. Colostrum's old-fashioned name is beest or beestings. I'm just another mammal with her nursling.

As my baby suckles, breathing softly and rapidly, it finally hits me. I did it! I'm a mother! I inspect the baby for the first time, feeling her doll-like fingers and toes, her slender thighs, her belly, her heft. She is a pale creature, cute, and intensely focused on my nipple. Her eye, the one I see as she nurses, is clear and luminous. She hardly blinks. Her skull is as delicate as an egg. The lips are very red. She breaks her latch and strains to regain it, snorting and snuffling, her crimson mouth opening and closing like a sea anemone, becoming louder and slightly frantic. With my help she latches on again. She flashes a private smile around the nipple. I clutch her to me, rock a little, and see the first blush of sunrise on the water.

At this moment I flash-forward to our decades together. How the years will fly by, how much time is passing already. Our first night together is nearly over. I begin to cry.

It's the oxytocin, surely. The baby's suckling sent a signal to my hypothalamus, the hormonal command-and-control center, which relayed the message to the pituitary gland in the brain, which responded by pumping out the hormone oxytocin in forty to fifty bursts a minute. Known as the "hug drug" for its soothing effects on the nervous system—my excuse for getting sappy—oxytocin also makes the uterus contract during nursing, as it did in labor. This helps shrink the womb down to its prepregnancy plum size sooner, at the price of mild cramps. Oxytocin also causes the mammary ducts to contract so that milk squirts out the fifteen or so openings in each nipple. This is the milk-flow reflex. I prefer its informal name: the let-down.

The let-down is also a boost. Oxytocin in the bloodstream is like a massage to the nervous system. It lowers our blood pressure. The bloom of blood warms our bosoms, opens our hearts. Oxytocin receptors flourish in our brains during pregnancy and especially right after birth. The hormone especially charms the amygdala, the emotional hot spot. We become more trusting and feel more strongly connected with others. We bond with our babies better. We may get a touch emotional. Our levels of oxytocin in the first trimester even predict how strong our bond will be with the baby after birth.

When we're very stressed, milk dries up because oxytocin is suppressed by the stress hormone cortisol. Otherwise, oxytocin soars after orgasm and when nursing. Milan Kundera described it in *The Unbearable Lightness of Being*: "When she first felt her son's groping mouth attach itself to her breast, a wave of sweet vibration thrilled deep inside and radiated to all parts of her body; it was similar to love, but it went beyond a lover's caress, it brought a great calm happiness, a great happy calm."

Really? I have to be honest, I'm starting to feel uncomfortable. I look at my darling nestled against me, still working my nipple. Her eyes are closed. I don't want to interrupt her dreamlike reverie. Then the morning nurse thrusts her big sunny face through the curtain, and we both startle.

"Marvelous that you're breast-feeding, hon," she booms. "How long you been at it?" Her hand hovers near a magic board where she will keep track of every feeding and every diaper change. What goes in and what goes out.

"Three hours," I say proudly.

The nurse looks me over in smirking astonishment. My right nipple resembles a chew toy. Cracked and bleeding, it will make me weep for weeks. Thus, I learn my first lesson as a mother: the baby was doing what's known as nonnutritive sucking, soothing herself, and it's murder on the nipples. Don't let your newborn do that.

Genes, hormones, instincts—we need them, but we can't depend on them to be a good parent. We're a funny species. Mothering doesn't always come naturally, and what constitutes natural is controversial.

During the postpartum months (the first three after giving birth), I'll need to navigate the nuances of breast feeding, mood swings and motherese, crying and comforting, sleeping, questions about my health, the baby's health, minding and my mind, now and in the future. I'll wonder where maternal instinct comes from and if I have it. My breasts will become more instrumental than ornamental, as the saying goes. It will take longer than I thought to heal; in the first month, even sitting down will be a challenge, and bowel movements will be excruciating. Relationships with other women will heal and become healthier. I'll tear up at the sappiest things. I'll relish poetry more than ever before. Everything will become tinged with new meaning and purpose. The Mommy Brain is not just born, it's also built. And it's always under construction.

> The moment a child is born, the mother is
> also born. She never existed before. The
> woman existed, but the mother, never. A
> mother is something absolutely new.
> —Osho, spiritual teacher

DO MOMMIES HAVE BETTER BRAINS?

Ten weeks have passed since my baby's birth, and I can tell a hunger cry from gas. I know how to trim doll-like nails. I wonder if I've lost my knack for a witty comeback, and I misplace my glasses and keys every day. I'm spending a lot of time with my mother, and childhood memories slam me by surprise. The idea of sex is unappealing. I've bought a book on infant massage. I've doubled a lentil soup recipe and frozen half of it. When we went on vacation, I forgot to bring my cat—my loyal, loving companion of twenty years—and had to turn around to get her. I can walk and talk (about babies) and nurse at the same time. I notice babies everywhere, and what brands of strollers they're in, and what toys are clipped onto their blankies or bunting, and whether they're with mommies or nannies, and, if mommies, what those mommies look like and how I compare.

At three months postpartum, my brain would fascinate neuroscientists. I happen to fall into what they think of as the sweet spot for the development of the Mommy Brain. This is the magical time when my brain should undergo a massive makeover and become more maternal. After months of shrinkage, I feel like a sagging model about to get a bust enhancement.

My inspiration is a study led by neuroscientist (and new mom) Pil-young Kim when she was at Yale University's Child Study Center. Kim and her colleagues recruited nineteen breast-feeding mothers and took MRI scans of their brains at two to four weeks postpartum and again at three to four months. Comparing the two sets, the researchers found in the three- to four-month scans a mother lode of new neuron-dense gray matter in areas where a new mom might need it most: planning and execution (prefrontal cortex), motivation and reward (amygdala and substantia nigra), perception and sensory integration (parietal lobe), and maternal instinct/hormones (hypothalamus). Women who took obvious pride and pleasure in their newborns showed the biggest boost in gray matter, whether they were first-time mothers or veterans. These were women who called their babies beautiful and perfect, and considered themselves "blessed." They have Mommy Brains.

An adult doesn't grow new gray matter unless a truly life-changing event has happened. Giving birth qualifies. No one told me about the ecstatic intensity of gazing into my baby's eyes. Nothing prepared me for how delectable my daughter would be—pushing her feet on my chest at the changing table, going pie-eyed over picture books, snuffling and gasping around the nipple, emitting a sweet milky essence from the soft spot on top of her head. Caring for a newborn is a sensory extravaganza.

The Mommy Brain needs what neuroscientists call "infant-related stimuli"—seeing, stroking, smelling, and (maybe) suckling—to trigger brain growth. The baby is the spark our neurons have been waiting for. Throughout pregnancy and labor, hormones—estradiol, prolactin, oxytocin, progesterone, and cortisol—have primed regions engaged in mothering, and these hormones will continue to shape our brains. (A radical theory is that fetal cells also trigger the growth; see page 220.) The new neurons have flourished in the areas related to smell sensitivity and in the hippocampus, the memory center. Everything is all teed up at the time of birth. After the newborn arrives, the sensory-processing parietal lobe is flooded with new sights, smells, and feelings. These sensations trigger the cascade of changes in the brain. Giving birth is not enough; we also need to *feel*.

The Mommy Brain is an addict's brain, and the newborn is the fix. She's candy. At six weeks, my baby started to smile. Her grins, gurgles, coos, and other cues stimulate the release of dopamine in the same reward circuits of the brain as cocaine. Kim and her team at Yale found

significant growth and activity in these motivation-and-pleasure circuits: the amygdala, substantia nigra, ventral tegmental area, and prefrontal cortices.

Addicts are usually bores, and new mothers are no exception. I can
babble on and on about how the baby babbles. The baby adorably bursts
into my every thought. Moms spend an average of fourteen hours a day
obsessing on their baby in the weeks after childbirth (fathers spend
seven). It's not obsessive-compulsive behavior, but it's in the same family. Evolution motivates us to find mothering this insanely rewarding.
The baby needs to be fed, pampered, protected, soothed, and stimulated. Much of this is groan-inducing tedium. Without an absolutely
smitten dopamine-rich superresponsive amygdala and substantia
nigra, how and why would a mother ever do what needs to be done?

For me, knowing this addictive feeling involves a sleight of neuro-
chemistry doesn't detract from the pleasure any more than knowing
that we're wired to love carbs detracts from my love of chocolate.

The Mommy Brain is also a less fearful and more aggressive brain—
at least when it comes to protecting the baby. Five days after I gave
birth, I was with my mother and the baby in a taxi. We were on our way
home from an appointment with a lactation specialist. It was a swel-
tering afternoon, and my newborn was a livid shade of red. I asked the
driver to turn up the air-conditioning, but he refused. Then he started
to drive erratically, gunning the gas and slamming the brake. His meter
ticked too quickly. "Driver," I said in a loudspeaker voice. "Pull over
now." The man twisted his neck around to peer at my bloated, postpartum face, and let out a crazed laugh. Something was off here, and I
didn't like that he didn't put his license on display as required by law.
"Now," I growled, holding my newborn close to my chest. "Or I'm going
to report you."

It worked; he swerved to the curb. "Pay me," he said. I refused. Once
we were safely on the street, my mother turned to me, astonished.
"You're like a mother bear!" It was true: don't harm my cub.

Not to diminish my heroism, but mother rodents get sassy and fear-
less too. Among mammals, motherhood appears to dampen the stress
response so that mothers can cope better with the anxiety of birth and

rearing a baby and not overrespond to stressors. The hypothalamus, command-and-control center of hormones, sends fewer frantic signals to the pituitary and adrenal glands (called the HPA axis) to produce stress hormones. Oxytocin and prolactin, which are abundant in us moms, suppress stress. Oxytocin also makes us mothers more aggressive when it comes to defending our young. I had just breast-fed my newborn before I attacked the taxi driver. I wonder if my hormonal state contributed to this uncharacteristically calm and courageous behavior. Similarly, mother mice strut out to the exposed arms of a maze, an unsafe area that normally makes them cringe. This is especially true when their pups are out there flapping in the breeze. Protecting babies is more of a motivator than fear is a deterrent.

The Mommy Brain is also a scheming, multitasking brain. From an evolutionary perspective, motherhood needs the prefrontal cortex (PFC) to be in on the act. The PFC helps us plan and project. It helps us choose between right and wrong, predict future events, be flexible, decide what's the most important thing at a given time, and exercise self-control. It makes good sense that this is another region where postpartum moms grow greater gray matter. Seeing the new growth on the brain scans of new moms, Kim and her colleagues suggest it's associated with their "increased repertoire of complex interactive behaviors with infants." The PFC is the mastermind behind my freezing soup for a future supper and tucking a pacifier in my bra before heading out. It's why my husband and I are planning a vacation weeks ahead, unheard of for us. It's why our fridge is full of food for the first time in our married life.

The Mommy Brain has a memory like a steel trap—or a sieve—depending on what's required of it. Peering into mothers' brains, deep in the memory center of the hippocampus, researchers have found new growth called dendritic spines. They help transmit information signals. Mother rats "beat virgin ass," in the words of neuroscientist Craig Kinsley, when it comes to remembering the location of a cache of sugary cereal or how to navigate a maze efficiently. For a rat, this is obviously useful in helping her forage for her brood and find her way home. Their spatial skills are vastly superior. Fascinatingly, the more times a female rat gives birth, the better her memory and the less risk of neurodegenerative disease—for the rest of her life. But for human mothers, the benefit is less clear-cut.

The mystery is that if our hippocampus, like that of mother rats, shows signs of new growth, why do we get momnesia? Why do we forget to pay the bills and return calls? Why do we miss appointments, misplace milk, and allow the name of a person we just met to slip through our memory filters? The studies that address maternal memory only add to the confusion. One found no difference in women's cognition at ten months and beyond when compared to childless women. Others find that women, even nine months after giving birth, continue to struggle with some types of tests: memorizing words, retrieving them, recalling what we said we are going to do. Words may be at the tips of our tongues. We find them, but we're slower at the draw. (I am certainly struggling more than usual to find words, but fatigue and stress may also be to blame.) The daily vitamin D pill becomes a weekly ritual, and the bottom of the teapot blackens from being forgotten on the burner.

Whether and why maternal memory is impaired is of intense interest. In the first few months we're often sleep deprived, losing up to seven hundred hours of sleep before our babies are a year old (although studies claim to factor in the toll of sleep loss). Some researchers blame the high levels of oxytocin in new mothers. Known to impair memory and attention and mellow us out when breast-feeding, the hormone helps keep the nursing mother fixated on the baby only. Try having a discussion about something technical while the baby's at the breast. I have, and it's like wading through quicksand.

I think it's safe to say our memory has not failed; it has focused. Mothers pay attention to what matters most: the baby's safety and well-being. My ears are so sharp now that I wake up in the middle of the night when my neighbor's infant cries. I might lose my glasses and keys, or stop tracking time, but my baby's okay. A mother's memory circuits are engaged in the mastery of hundreds of new skills: how to burp the baby, how to read the baby's face, how to nurse, how to install a car seat. Our multitasking PFC is in overdrive; perhaps we strain ourselves too much to remember and anticipate the small stuff. Does it really matter if I don't keep track of every birthday anymore, or if Google becomes Mommy's little helper for names and facts? As long as we don't throw our baby out with the bathwater, who cares if a few bits go down the drain?

Standardized psychology tests cannot fully grasp the ways our brains have improved. We're undergoing extreme on-the-job training. We're singers and soothers and degassers. We're mind readers, mediators, housekeepers, party planners, puppeteers, pillows, preachers, chefs, cows, and medicine women. Much of this is new to us, and any form of learning changes the brain. Neurons that fire together wire together. New neural networks emerge in pregnancy and motherhood, and experience strengthens them. Our gray matter becomes very colorful.

What isn't black and white are the answers to many of our questions. How much does the Mommy Brain vary among women? Can moms who adopt or hire a surrogate experience some of the same brain gains without pregnancy hormones? Could brain scans help identify women at risk of postpartum depression or child abuse? What about moms who love their babies but don't exactly find them addictive? How long might a brain boost last anyway? And is there a Grandma Brain? Watching my mother soothe my colicky newborn, I'd bet that maternal circuits, even if they have been asleep for a couple of decades, are able to reawaken. The Mommy Brain may be with us for the rest of our lives.

There's a Daddy Brain, too, although the brain gains (and losses) that a man experiences are not as dramatic. Preliminary studies have found that fathers undergo some of the same hormonal changes that women do and that men who have more than one child have especially high levels of prolactin. Based on her data, Kim has a hunch they also have new growth in limited regions such as the prefrontal cortex. Their amygdalas show the same activation as mothers do when hearing a recording of a baby cry. Another study found that the higher fathers score on tests of parental care, the more gray matter they have in their left prefrontal cortex.

All in all, is the Mommy Brain a better brain? It certainly excels at being a mommy. "Comparisons are odious," my husband reminds me (because I forget). Sometimes I think I have a better brain now; I'm braver, more grounded, more sociable, and more empathetic than when I was childless. Sometimes I think I prefer my mind when I had the capacity or leisure for laser-sharp focus and abstraction. I cannot decide which brain is better. As the Shakespearean saying goes, "There is nothing either good or bad, but thinking makes it so."

At least I'm still thinking.

Why My Baby Is Cutest

I now know the secret behind the survival of our species: babies' smiles. Most infants start to social smile within the postpartum period, the first three months. A flash of their gummy grins carries us through the colic, fatigue, sacrifice, resentment, and all the other reasons that a woman might regret reproducing. The smile is really *that* rewarding, *that* motivating, *that* addictive. It says, "I remember you, thank you, I love you." As I write, my computer wallpaper features a photo of my daughter at three weeks, looking up at me brightly, her mouth set in a newborn's eager overbite. Seeing this, my heart surges.

Curious about this magic behind maternal motivation, neuroscientists have recruited new moms and scanned their brains using functional magnetic resonance imaging. Taken together, these studies involve showing each woman photos or video clips she has not seen before—a child who is unknown or a friend's child, smiling or not, and others of her own child. Any cute baby face evokes some sort of activity in a mother's brain. But *only one's own smiling baby* stirs up the dopamine-driven reward-and-motivational circuits (ventral striatum, substantia nigra, thalamus, caudate nucleus, and nucleus accumbens) and areas associated with attachment (especially the orbitofrontal cortex). There is nothing more rewarding and addictive to most new mothers, especially in the postpartum months, than their baby's smile. Not even chocolate.

Researchers have also discovered another fascinating feature of baby mind control: we overlook our own baby's flaws. Gazing at our own child's smiling face can shut down areas of the brain associated with social judgment, negative emotions, and critical thoughts (medial prefrontal, inferior parietal and posterior cingulate cortices, and amygdala). This love-induced blackout must be behind why 73 percent of mothers (and 66 percent of fathers) rate their baby as "excellent" in his first three months, even when the child is screaming his head off half the time. If my baby had a face only a mother could love, I'd never know it. The same judgment circuit in the brain shorts out when we look at photos of a lover or partner. Love makes us blind to anything we don't want to see. So all we see is the smile.

IS MATERNAL INSTINCT SHAPED IN INFANCY?

I'll never forget the moment I first heard about the "wet cement" phase of infancy. This is when babies form their impressions of how nurturing their world is, and these impressions forge their emotional lives: how confident, balanced, and loving they will grow up to be, and how nurturing they will be when they themselves become parents. At the time I was in a psychobiology class in college. Motherhood was light-years away, but I found myself gazing out the window at a misty fall day and reflecting on what type of parent I'd be someday.

As I write about the topic of maternal instinct, my ten-week-old baby sits on my lap. She distracts me as I peck at the keyboard, but never mind. How could I write about mothering if, five feet away in a swing, the "mechanical nanny," I feel two burning eyes on me?

"Patience, little one," I murmur. The baby scratches my breast.

Scientists have long known that when baby rats are separated from their mother for hours daily, they grow up to be abusive, apathetic moms. Neglected infant monkeys too are sad sacks; they're frazzled, subordinate, jumpy, clingy, and depressed. They too become deadbeat mothers. The meanest thing you can do to baby mammals is not to touch or love them. We're warm-hearted creatures.

But here's the thing. If you separate baby rats from their moms for fifteen minutes a day, during days 5 to 10 of their lives, as neuroscientists Michael Meaney, Ian Weaver, and Moshe Szyf at McGill University did, and then reunite them with their moms afterward, those babies become calmer, more adventurous, and more nurturing than other rats.

Alas for all us tired new moms, it's not the fifteen minutes of daily separation that does the trick. It's what rat mothers do when they get their pups back: they become supernurturing. Awash in some rodent version of maternal relief, mothers lick and groom with gusto. They bend over backward, literally, to give their brood access to a maximum amount of breast milk. Their pups grow up to be nurturing mothers. How they were reared as infants shaped their maternal instinct.

"I love you, sweetie!" I say to my daughter reassuringly and turn back to the screen.

Scientists are just starting to figure out how good mothering influences our genes and hormones early in life. We know the process depends in part on a group of hormones called glucocorticoids. Among other duties, glucocorticoids influence our stress response. We have receptors in our brains for them, including in the hippocampus, which directs memory and attention. Glucocorticoid receptors are like antennas for what's going on in the body, and the more we have of them, the more we can tolerate stress. They send signals back to the adrenals, effectively telling them to switch off stress hormone production. The more antennas (receptors) we have, the faster we can shut down the stress response. Baby mice raised by nurturing June Cleaver moms have so many glucocorticoid receptors that their hippocampuses are plush with them. These nurtured pups grow up to become relaxed, adventurous, attentive moms. The opposite happens for babies with lax and loathing mothers. They have a threadbare glucocorticoid receptor count, and grow up to be lax and loathing, just like their moms.

Now, the fascinating part. There's nothing fated and unchangeable about the number of glucocorticoid receptors we get. Whether we or our children have many or few depends on the way our genes express themselves, and that depends on the environment we are thrust into in our early years. It's epigenetic. Like knobs on a tap, the genes for glucocorticoid receptors can be "turned down," or silenced (called methylation), or "turned up" (demethylation).

We come into the world with the genes for glucocorticoid receptors turned down (methylated), which would give us a low tolerance for stress. This sounds bad, but it isn't necessarily. In a stressful, dangerous, motherless setting, a person or animal might benefit from being anxious, cautious, and focused on self-survival. But if babies receive maternal love early in life—a sign the world they are entering is good and stable—their genes demethylate, or increase, production of the receptors. The result is a mammal that's friendlier, more confident, and secure, and can give and receive affection. Scientists think the process may be triggered by the soothing hormone serotonin that is released in a baby's hippocampus as her doting mother touches her.

Sadly, one of the ways we know that lack of maternal care hurts human babies in the same way as rat pups is by looking at the brains of suicide victims who have had a history of child abuse. Compared to people who had relatively normal childhoods, suicide victims have significantly fewer glucocorticoid receptors in their hippocampuses. Orphaned Romanian infants in the 1980s who were given formula to drink but no touch or interaction were found to have sky-high stress hormones and difficulty giving or receiving love—even if they were adopted as young children by nurturing, supportive parents. (The glucocorticoid receptor gene is not the only DNA that helps decide whether we'll be damning or doting moms, but it's among the most interesting and proven. The McGill team identified a staggering number of genes— more than nine hundred—regulated by maternal care, including genes that process neurotransmitters oxytocin and serotonin.)

The most impressionable time for humans may be the first three years of life. This, according to developmental psychologist Mary Ainsworth, is when we develop our relationship with our mother (or primary caregiver)—the mother of all future relationships. Astonishingly, Ainsworth found that a person's attachment type in infancy is in many ways more predictive of future behavior than temperament. Babies whose mothers are neglectful or erratic, stingy or flaky in their affection grow up to be anxious, fearful, clingy or distant, or risk averse, and they have insecure relationships. Mothers who are sensitive to their infant's subtle signals (for example, perceiving hunger by a tiny mouth movement and knowing when the baby is really done feeding—not easy), respond to their demands (say, perceiving that raised hands are a plea for attention), and cooperate with their baby rather than interfering, tend to raise "securely attached" children. These kids are more likely to be kind and confident and form strong, warm bonds with others, including, someday, their own kids.

Secure childhood attachments even show up in the structure of the brain. A recent study at Yale University found that women who had nurturing mothers show more activity and gray matter volume in brain areas that help them understand their own children's intentions and mental states (hippocampus, fusiform gyrus, parietal lobes, middle frontal cortex and other regions). Mothers who suffered neglect in infancy have amygdalas that are more sensitive to negative emotional stimuli, making them interpret neutral situations like a baby's cries overly negatively. Good parenting isn't necessarily in the genes; it's in the environment that affects the ways genes behave.

Will we someday have gene-targeted therapies that take us back to the "wet cement" window so our genes might act as if we grew up cuddled and cosseted by adoring mothers? There have been promising animal experiments that have wiped out some of the side effects of maternal mistreatment. One found that among young rats suffering from maternal neglect, those that are placed in an enriched environment—with exercise wheels, toys, and an active social life—become more attentive mothers than those in less stimulating settings.

What does all this mean for us? Decisions we make in our lives may help us overcome some of the backlash of bad babyhoods, and thus become better parents. This includes talk therapy, spirituality, self-awareness, or any other form of willful personal transformation. Humans are complex, and so is gene behavior. Among us are those who crawl naked through rubber birth canals, to be pulled out by patchouli-wearing friends who'll cradle them skin-to-skin. Are these reborns resetting their glucocorticoid receptors too?

Ainsworth, a childless professor, was attacked by feminists because her work could be used to goad and guilt women into staying at home for their child's first three years of life. We moms are under so much pressure already. Ainsworth responded evenhand-edly. "Had I myself had the children for whom I vainly longed, I like to believe I could have arrived at some satisfactory combination of mothering and a career." There's nothing in the research that pre-vents us from saying another loving caregiver or a stable group of adults could provide the same benefits. I believe the best legacy I could give my infant daughter—and her children if she chooses to have them—is love and nurturing, and that comes in many forms. It still takes a village to raise a child.

Alas, the mechanical nanny doesn't qualify. I glance at the fussy baby on my lap, only slightly sated by the breast I tugged out. Seeing me see her, she breaks into a beatific nose-crinkling smile. "I l-o-o-o-o-ve yo-o-o-o!" I gush. Her first impressions are still forming. My heart soft-ens, my resolve hardens. Time for a break.

> Whenever I held my newborn baby in my arms, I used to think that what I said and did to him could have an influence not only on him but on all whom he met, not only for a day or a month or a year, but for all eternity.
>
> —Rose Kennedy

WHY DO WE SPEAK MOTHERESE?

It's amazing what hidden instincts and talents surface in the first weeks of motherhood: the ability to recall old nursery rhymes, the use of the hips, the knack to catnap. But by far the most disturbing, at least to me, is fluency in "motherese," that distinctly grating form of slow, simple, repetitive speech spoken with exaggerated vowels in a sear-ingly high pitch. No one ever called me the motherly type. How did I get this nut in my tool kit?

"P-o-o-o-p!" I croon to my three-month-old daughter during her diaper change. "W-o-o-o-w! Wh-o-o-o-a! S-o-o-o-o much p-o-o-o-op!" My voice escalates as I reach the last word, and even to my own ears I sound shrill. My baby, who had been dreamily gazing at the ceiling fan, turns and gives me an ecstatic grin. "Fa-a-a-a-n!" I screech. "Ye-e-e-s, we lo-o-o-ve the fa-a-a-a-n!"

Good lord. "Motherese" or any other word for it—"baby talk," "parentese," or, in scientific circles, "infant-directed speech"—feels right, but I'm shocked every time it eases out of my mouth. Just three months ago, I complained how irritating it is, how condescending and undignified. An infant, I believed, should be exposed to proper speech, spoken at a regular rhythm, in a normal register. But here I am baby-talking with impunity.

I offer three reasons that this is so.

The first argument is that motherese attracts the baby's attention. I've tested this on my daughter, and I confess it's true. "Have you pooped again, sweetie?" spoken in my normal ironical tone does nothing. Even louder with a smile, nothing. She stares wistfully at the fan or punches the air. Then I try, "Did we-e-e-e po-o-o-op?" Riveted, the baby turns her head and examines me closely. A fist is raised triumphantly, and her eyes are locked on my face. Sometime after their first month, infants suck on pacifiers harder or look at visual displays longer, signs of interest, when they hear motherese. Truth is, infants adore melodic, rhythmic, high-pitched expressiveness. Speak motherese in a dull, flat monotone as depressed parents do, and the words fall on deaf ears. That friendly familiar singsong that babies know is targeted at them helps regulate their emotions. Motherese is literally music to a baby's ears.

The second reason I'm a proud linguist in motherese is that babies exposed to it learn language faster and more easily, and they develop a wider vocabulary. (So, incidentally, do adults when learning a new tongue. When people from nonliterate cultures hear an English sentence spoken in motherese, they can guess the intention much more often than when the sentence is spoken in an adult tone.) While the piercing pitch perks up ears, the lo-o-o-o-o-ong, slo-o-o-ow, and stre-e-e-tched-out mouthing of words helps them pick out vowels—the difference between *pee* and *poo,* for instance.

Motherese also makes it easier to tell where one word ends and another begins. To ears that don't know any better, "Did we poo?" can be, "Di dweepoo," "didweepoo," or "did weepoo." A hyped-up huge-eyed "Did we-e-e-e po-o-o-o-o?" highlights individual words like jewels. Each is buffered by pauses before and after, and shined by happy affect. When experimenters exposed infants to sentences in which the speech flowed in unexpected places, the babies knew something was off—but only if the words were spoken in motherese, not in solemn grown-up tones. Speakers of motherese put the most important word at the end of a sentence, which also makes it stand out. More than once during diaper change, I have seen my baby laboriously cooing o-o-o-s. I fear the child's first word.

The third reason I'm speaking motherese is that I have no choice: I'm hardwired to do it. I say this with tongue-in-cheek, but it's true. Everywhere on the planet people talk to babies in a squeaky, breathy, singsongy mode, even in indigenous cultures innocent of *Sesame Street* and other media. This suggests that motherese is at least partly instinctive. Mothers speak motherese, of course—and the higher our oxytocin levels, the more likely we are to speak it—but so do fathers, grandparents, some nonparents, and kids. Parents unconsciously amp up the motherese when their baby is three to four months old and just learning the sounds of speech, and slowly lapse into adult speak as other modes of language learning take over.

Motherese may even be the mother of music and language, according to one theory. About 4 million years ago when our ancestors became bipedal, infants could not cling to their parents as other primates do. This meant parents often set babies down while foraging or processing food. Face-to-face instead of flesh-to-flesh, mothers used language to entertain, soothe, command, or prevent infants from harming themselves. Over the millennia, motherese—once expressed in looping hoots and howls, fish mouths and saucer eyes—spun out into more complex rhythms and sounds and facial cues. Humans who were good at language and music were more likely to survive and find mates. Language-enhancing genes spread far and wide.

So here we are, millions of years later, and motherese is still our first language. But evolution can happen right before our eyes. Someday I'll ask my daughter, "Did w-e-e-e-e-e p-o-o-o-o-op!?" And she'll say, "Mom, stop being so scatological."

> You know how Adam with good reason,
> For eating apples out of season,
> Was "cursed." But that is all symbolic:
> The truth is,
> Adam had the colic.
>
> —Ambrose Bierce

WHY ARE WE STRICKEN WITH COLIC AND DEPRESSION?

Things have suddenly taken a turn for the worse. The baby has been crying for eight hours nonstop. She has been rocked and bounced, shushed and swaddled—with increasing force and desperation. She's not sick, not hungry. The diaper is dry. Earlier in the day, when lifted up to Mommy's face, eyeball to eyeball, she widened her eyes and cried harder. Now it's four o'clock in the morning, and Mommy is losing it. We're in the street in front of our home. I've wrestled her into a sling and am jumping up and down under the streetlight, singing "Amazing Grace" in agitated bursts.

The colic makes me wonder if I can ignore the baby. Perhaps I could put her down on the dewy grass and let her scream at the stars and the moon while I drop my head in my hands and weep. How sweet the sound.

Colic and depression are the two plagues that afflict healthy parents and their newborns. Both are astonishingly common. One in four newborns has colic, defined as a siren-like wail that lasts at least three hours straight, more than three days a week (usually in the late afternoon and evening hours), for more than three weeks, for no apparent reason. Postpartum depression is something like an adult colic. For at least two weeks, and usually up to four months, moms and dads who suffer from it feel despair and emptiness, fatigue, insomnia, and even thoughts of death. About one in five women suffer from it (and one in ten men, who often have it longer). While a baby can have colic without having a depressed mom and a mom can be depressed even if her baby doesn't have colic, the two are like evil twins. Baby's colic feeds Mom's depression. Mom's depression feeds Baby's colic. Everyone feels empty.

Why would Nature throw ice water on parents when we need to be our warmest? To help solve the mystery, evolutionary psychologists and anthropologists begin with a basic fact: babies are born with their fuel tanks near-empty. If they don't get food and attention swiftly and consistently, they stop running. The most effective way to get what they need is to cry, and the period of the most intense crying is the first three months, peaking at six weeks, and intensifying in the witching hours of late afternoon and early evening. Fussy babies follow this pattern in every culture around the world. Other baby mammals do too.

Incessant, inconsolable crybabies like my daughter (and me, when I was an infant, according to my mom) have an additional agenda, according to biologist Virpi Lummaa and her colleagues. The point of colic is to deceive. Even if the colicky baby is stuffed and spoiled, she demands more. Exploiting an adult's hardwired motivation to do anything to get a baby to shut up, these squeaky wheels often receive more parental grease. The baby doesn't get colic consciously, of course. It's in her code. Long ago, fussy babies may have had such an advantage over their calmer peers that genes related to colic spread. In one study of the Masai, tribal herders in Africa, babies with colic symptoms were more likely than easy babies to survive a drought. The parents couldn't ignore a flood of tears. (Happily, colic is not a predictor of temperament later in life.)

As manipulative as colic may be, it could also be an honest signal that the baby is fit and worthy of the extra resources. Observing my crimson-faced, fist-clenching, back-arching daughter, a bystander remarked, "Strong lungs, eh?" Aha, but that's the point. According to Lummaa's theory, colic is a signal to parents that the child is "high quality." All that angst may be up to twenty times more taxing than resting angelically. If a newborn weren't vigorous enough, our ancestors were likely to cut their losses in hard times by withdrawing food and attention, abandoning, or even deliberately killing the baby. There is no clinical evidence that colicky babies are actually healthier. If they are, it might be obvious only under the toughest circumstances: famines and floods, diseases, the loss of caregivers, and other hard knocks.

At four in the morning, the physically and spiritually drained mom of an inconsolable infant might not agree that colic prevents infanticide. Fact is, women with supremely fussy babies are more than twice as likely to have postpartum depression as women whose babies cry less. Some sufferers confess they secretly believe the newborn is a parasite or an anti-Christ, à la Rosemary's Baby. Some, like me, contemplate laying the baby on the grass to let her cry it out. Others fantasize about hurting the baby. Sadly, some do.

Except when they get help. Which brings us to why we get postpartum depression. Edward Hagen, an anthropologist at University of California, Santa Barbara, sees it as a labor strike—a mother's silent cry for help. Postpartum sufferers often lack support. Compared to other moms, they are more likely to be poor, debilitated or traumatized by the delivery, or in the midst of marital mayhem. The tragic and often unspoken truth is, 90 percent of parents experience a decrease in marital satisfaction in the first months after a baby is born, and many have a moderate to severe crisis. The depression and dissatisfaction often begin with the mom and spread to the dad. A man is more likely to "catch" postpartum depression if his wife has it.

In the distant past, a severely depressed woman might have decided that the costs of the birth outweigh its benefits and abandon the baby. Infants may have occasionally died, but it's more likely that the mother's shutdown drew attention from others in her community. We evolved in small, closely knit groups where kin helped care for an infant. The !Kung, a hunter-gatherer tribe in Africa, still embrace this way of life. Sure, !Kung babies get colicky, but the bouts are shorter because babies are breast-fed continuously, several times an hour, and the most demanding ones can revel in more maternal care than any one woman could handle (infants cry fifteen seconds or less before someone in the extended family, including kids, rushes to take care of them). !Kung moms get postpartum depression too, but when they do, others take on more of the workload, avoiding psychological damage inflicted on the baby by a mom who (at the moment) can't love.

The evolutionary explanations are credible and appealing. But they may not be the whole answer, at least not for everyone and not all the time. For some babies, colic may be a manifestation of an immature nervous system, a developmental reorganization, digestive woes, gut bacteria, or air in the intestines. I have talked to doulas who swear colic is the first screaming sign of a dairy or soy allergy. For some moms (and dads), postpartum depression may come down to stress, fatigue, plummeting hormones, or low thyroid levels. A provocative theory from the evolutionary psychologist Gordon Gallup is that bottle feeding, an option available only in the past hundred years, leads to postpartum depression because lack of lactation has the same hormonal effect as the death of the infant. Truth is, colic and depression are such general conditions that there's room for more than one explanation.

I've not descended to the depths of depression. But my baby has had enough soul-numbing nights of colicky crying for me to understand it. At these times, my husband helps out more. And when he reaches his threshold, my mother has stepped in, so I am not the only wretch singing and swearing under the stars.

It really does take a village to raise a child. It also takes Amazing Grace.

Why We're Lefties When We Cradle

By observing the way they hold their newborns, researchers can predict, with middling accuracy, which women have the highest risk of postpartum depression. Stressed and depressed moms are the ones awkwardly cradling their babies to the right. Most girls, women, and fathers—left- and right-handed alike—cradle a baby in their left arm instead. (And so do chimps, gorillas, and orangutans.) We do this naturally and unconsciously.

Researchers don't know for sure why we have a left-cradling instinct. Some believe that left-cradling places babies near the heart, which soothes them (although even women with hearts on the right tend to cradle their babies to the left). The most popular theory, an intriguing one, involves the way our brains (and those of other large primates) process emotional information. When a baby is cradled to the left, the right half of her face is partially hidden, but the left half is not. It turns out that the left half of the face is more emotive; it reveals cries, coos, moues, and other cues more strongly than the right side. It's in the baby's best interest to show her caregiver more left face and also for the caregiver to show the baby more left face. Just as important, the caregiver's left eye and left ear are closer to the baby. This is important because information coming from the left eye and ear go to the right hemisphere of the brain, which perceives emotions more effectively than the left.

Who would subconsciously shut out a baby's cries and grimaces? Depressed and stressed-out people might—which could explain why they are more likely to cradle babies to the right. These mothers are more likely to interpret their baby's behavior the wrong way—for instance, a cry as a sign of defiance and criticism rather than a normal thing that babies do. Even temporary stress might make us turn the other cheek.

WHAT'S LIVING IN OUR MILK (AND WHY)?

Not long ago, the chef at my local brasserie made a canapé of breast-milk cheese with figs and Hungarian pepper. The milk he used in the recipe was locally sourced—from his wife, a nursing mother. Pleased with what he had done, the chef used his wife's breast cheese in a dish with truffle dust and burned onion chutney, with a light soy sauce, alongside crackers, and as a topping for pasta. He was rumored to be working on a breast-milk gelato. The public was repulsed and outraged, but to the shock of those who tried it, human milk tastes refreshingly light and sweet. It's like the milk at the bottom of the cereal bowl. Connoisseurs deem breast milk cheese mild and hardly inedible.

A spoonful of sugar really does make the medicine go down, for breast milk is actually an immune system booster in disguise. It's a White Knight hiding in a birthday cake. "Lactivists" call it the White Blood because 1 million killer white blood cells swarm in every drop. Like red blood, it contains hormones, immune factors, vitamins, and minerals. No image is more convincing to a woman hedging about whether she should breast- or formula-feed than the sight of the respective fluids under a microscope. Breast milk resembles life itself: crowded and chaotic. Formula looks like a wasteland: a flat landscape pocked by the occasional air bubble.

Nursing my daughter, I'm colonizing her with life forms—living white blood cells. They come in three flavors: macrophages, lymphocytes, and neutrophils. In the first ten days of a baby's life, there are more of these fighters in his mother's breast milk than in blood. Macrophages ("big eaters") are a type of white blood cell that engulfs foreign invaders and tags them so the other immune system workers recognize them. Lymphocytes, known as B cells and T cells, manufacture antibodies and blow up infected cells, respectively. They help activate the infant's own immune system and encourage it to mature faster. The thymus, the spongy organ behind the breastbone where T cells are made, is twice as big in breast-fed four-month-olds as in the formula-fed. Neutrophils, the third type of white blood cells, release a fibrous web that traps and kills microbes. Then, like the vicious hairy spiders they resemble, they release superoxide, a toxic bath of enzymes that converts to chlorine bleach.

These assassins have an arsenal of antibodies—IgA, IgG, IgM, IgD and IgE—that are also found in breast milk. Secreted by B cells, antibodies are not alive in the way white blood cells are. They're alive in the way memory is alive. They are the molecular memory of infectious viruses and microbes we've encountered. Every time we've beaten or been beaten by one of these invaders (called antigens), our antibodies remember its "face" in case it dares show it again. Antibodies are precision killers; they know to spare the good-guy bacteria in the gut that keep the hordes of bad-guy bacteria at bay. If we get the flu while breast-feeding, the antibodies we produce will protect our babies.

Everyone in the family may be sick except the tiny breast-fed cherub, who is rosy cheeked and well. In breast milk, we pass along a living legacy of defensive strategies, the intelligence of battles lost and won, with the hope that our babies will also benefit, at least temporarily, from hardships we've experienced in our own lives.

Other special agents in breast milk also have a role: lysozyme (promotes growth of healthy flora and is anti-inflammatory), lactoferrin (binds iron that bad-guy bacteria feed on), bifidus and lactobacillus (good-guy bacteria that kill their bad-guy brethren with acid), oligosaccharides (nondigestible substances that feed the good-guy bacteria), and lactose (the main carbohydrate source, as well as a promoter of good bacteria). German researchers studying breast milk found that it contains a compound known as HAMLET (human alpha-lactalbumin made lethal to tumor cells), which is now being researched as a cancer cure.

The immune forces of breast milk specialize in guarding Baby's perimeters—the mucous membranes, and in particular the leaky gastrointestinal tract where invaders creep in. Breast-fed babies are ten times less likely to be afflicted with infectious diarrhea because the walls of their gut are literally whitewashed with antibacterials (against *E. coli*, salmonellae, shigellae, streptococci, staphylococci, pneumococci, poliovirus, rotaviruses, and so on) in the milk. You can leave a bottle of the human milk on the counter for hours and it won't go bad. An ancient remedy for eye infection is breast milk, raw. Like Lysol, it kills on contact.

In a way, the breast is just picking up where the placenta left off. During pregnancy, the placenta passes antibodies directly to the baby from our bloodstream. After birth, breast milk provides an extra boost, an influx of new troops to the GI tract. They remain in the baby's circulation for months. The bigger the boost in the beginning, the better.

It takes several years for a baby's immune system to mature. No one knows why babies are born with such undeveloped immune systems. It may be a biological tactic to help avoid rejection by the mother during pregnancy. Or it may be that babies must focus all their energy on brain building, not developing a first-rate defense. Babies need milk anyway, so Mother Nature, ever resourceful, simply packs it with an extra punch.

Of course, a question pops into the mind of every nursing mother at three in the morning: Why can't Daisy take over? Cow's milk has the same ratio of fat as human milk. Raw, it also contains live immune factors, the inspiration for the grassroots movement for unpasteurized milk. But the problem, it turns out, is that cow's milk has more protein and salt than human milk does. This sounds good, but it's not. Baby kidneys can't tolerate so much salt, and baby stomachs can't digest the clunky protein, called casein, that makes milk curdle. (A good cheese needs casein, which is why a chef needs to add a little cow's milk to a breast milk recipe.) Give too much casein to babies, and they'll cramp and explode with diarrhea. In contrast, the protein in human milk, which is only 40 percent casein (and 60 percent watery whey), is digested so efficiently that a breast-fed baby's bowel movements hardly smell, whereas formula-fed poops are famously foul. You're smelling all that undigested protein.

Breast milk is better than any udder milk, as the saying goes. The human fuel has the perfect combination of proteins, vitamins, carbs, and brain-building fats. Breast-fed babies have a decreased chance of obesity because they learn to regulate their own intake of calories rather than being forced to empty a bottle, and the hormones in it (leptin, ghrelin, and obestatin among others) may influence the programming of energy balance regulation. Whether this benefit lasts into adulthood is unclear.

What is clear is that breast milk primes the immune system. It reduces inflammation. It trains the body not to attack itself and to permit the flourishing of good, protective bacteria. This translates into other long-term benefits: a reduced chance of juvenile diabetes or rheumatoid arthritis, Crohn's disease, celiac disease, and colitis. Mysterious and magical, breast milk gives us immunity, and scientists have barely skimmed the surface of how it works. The stuff is alive, it's the elixir of life, it's holy water, manna of the mammaries.

The World Health Organization guidelines recommend breast feeding exclusively for six months, followed by two years or more of breast feeding with supplementary solid food. Some anthropologists claim our species is meant to nurse for up to seven years, the age when some kids are launching their first dot.com. Nursing eighty minutes a day, spread between at least six feedings, may suppress ovulation for up to a year and a half and act as a natural, albeit unreliable, form of family planning. It diverts about five hundred calories of a mother's daily diet to her baby, helping us take off the baby fat.

"Aim to nurse for one year," advises my baby's pediatrician, referring to the American Academy of Pediatrics guidelines. Some studies show incremental gains in immunity beyond baby's first birthday, although in countries with good hygiene, the benefit is not as obvious.

I am determined to make the one-year mark. I move to put my right hand over my left breast, over my heart, and then pause. "But what if I can't?" The doctor nods seriously. "Nurse for as long as you feel able," he says.

Truth is, breast feeding is like sex; it's beautiful and pleasurable when done freely, naturally, and when the juices flow. It can be agony otherwise. Get stressed or sick, and the mammaries struggle. For better or for worse, you and your baby are still a unit for as long as you're nursing. Soon after the salads arrived at my first postpartum business lunch, my milk-engorged breasts tingled and I sensed that at home, all the way across the city, my hungry baby was howling. While 64 percent of American moms breast-feed in the first month, fewer than half of us make it to the six-month mark. Only 16 percent of us will nurse a one-year-old.

Three months of breast feeding (or pumping) is better than one month, six is better than three, a year better than six. Three months into it now, I can say from experience that it's a real commitment to my baby and her health. But it can also be pleasurable for me too; an opportunity to cradle her heartbreakingly defenseless little head and pour my strength into her. For when breast feeding is sweet, it is very, very sweet.

> Breast milk: the gift that keeps on giving.
> —Anonymous

ARE BREAST-FED BABIES REALLY BRAINIER?

Kids who were raised on mother's milk score two to seven points higher on IQ tests than their formula-fed peers. Not one study shows formula-fed kids scoring higher on intelligence tests than breast-fed kids. This much is fact, and it's on this basis alone that I've joked with my husband, a formula-fed baby, about his early shortcoming in life. "You could've been a contender," I say, affecting the arrogance of a breast-fed brat. But the joke's on me, because the extent to which breast feeding affects the brain is actually a topic of hot dispute. While the immune-system boosting benefits of breast milk are clear, its brain-boosting benefits are somewhat . . . well, milky.

Here's the rub: the moms who breast-feed tend to have higher IQ scores themselves. Compared to moms who formula-feed, they're more likely to be better educated and from a higher socioeconomic class. This means that nurslings might be smarter than formula feeders because of their mom's genes or influence—not the milk.

One way researchers could try to control for this is to take a group of moms with the same background, education, and IQ levels and assign half their babies the breast and the other half the bottle. But this would be unethical. So they've done what they can: factor in maternal IQ to try to account for the difference. When they do this, the data look a lot different. In many, but not all, of the studies, the difference in IQ between breast- and formula-fed kids disappears.

This hasn't soured all breast milk researchers. Of the studies that claim to adjust for maternal intelligence and other factors, some find a persistent IQ advantage—three points or more—among kids who were breast-fed as babies. One recent Belarusian study of more than seventeen thousand kids, followed until age six, found a six-point difference in full-scale IQ after taking all factors into account. Several studies also found that the duration of breast feeding is important: the longer a baby is breast-fed, the higher the IQ, with a couple of points higher every additional season, capping at nine months. One research team found that the babies who chugged the most breast milk had a larger brain circumference (volume) at nine months, which is related to better mental performance at age nine. A long-term study found a minor advantage of breast milk consumption for adults younger than thirty, but another study on senior citizen males did not.

Many of the well-researched studies that found an IQ advantage to breast milk suggest that the secret ingredients are docosahexaenoic acid (DHA) and arachidonic acid (AA). Known as omega-3 long-chain polyunsaturated fatty acids, these compounds in breast milk migrate to the brain, where they end up providing structural support for cell membranes. Omega-3 fatty acids are abundant only in breast milk, not in cow's milk or infant formula (although some are now supplemented with DHA).

Is it the DHA and other omega-3s in breast milk that really make an IQ difference? It may depend on how much of these fatty acids mothers eat in their diets. Fish lovers have up to twenty times more omega-3s in their breast milk than diehard meat-and-potato types. One way to test whether DHA is so special is to prescribe breast-feeding women DHA pills, restrict their diet, and compare their results to those of women who don't take the pills or eat fish. One research team split their volunteers into groups that received different dosages of DHA supplements for the first twelve weeks postpartum. Babies whose moms had the most DHA at the end of the study scored highest on a test of mental development at one year old but not at two years old, presumably because other factors kicked in.

Many studies are only shakily in support of maternal consumption of DHA. One found a positive association between DHA and IQ for girls at nine months but not at four months. A research group at Baylor College of Medicine found that two-and-a-half-year-old toddlers whose moms took two hundred milligrams of DHA when breast feeding had significantly better psychomotor skills than those whose breast-feeding moms didn't take the supplement. At five years old, they had better sustained attention skills. Oddly, no differences showed up when they were younger, perhaps because these abilities don't go "online" until later. Mothers in Norway took either twelve hundred milligrams of DHA in a cod liver pill or a corn oil pill every day from eighteen weeks of gestation through three months postpartum. The preschoolers of the DHA pill poppers scored higher on a developmental test, but it's impossible to tell whether the boost was pre- or postnatal.

Only a person's IQ scores in mid-childhood, around nine years old, predict how he'll score as an adult. This means the jury is still out on any long-term cognitive benefits of omega-3s in breast milk (or added to formula) because they haven't been studied for decades. Any brain-boosting benefits of breast milk, with or without a fish diet or supplement, may well become diluted over time.

Clouding the water even more is research that finds that breast-fed babies are brighter *only* if they have the genes to process omega-3s effectively. Researchers at King's College London targeted a gene called FADS2, which is involved in metabolizing fish oil and other fatty acids. According to their study, if a baby has the common C variant of this gene, breast milk will make her smarter than unsupplemented formula would and by a staggering amount: nearly seven IQ points on average! About 90 percent of us have the C variant. If our nursling happens to have the less common G variant of the gene, no amount of breast milk will budge her IQ. This is a fascinating example of how genes interact with environment. Unfortunately, a recent follow-up study could not confirm it; it found instead that kids with the less common G variant are the ones who get the brain boost from breast milk.

The IQ benefits of breast milk are much more obvious for preterm infants because their brains are tremendously sensitive to nutritional influences. One study of nine-year-olds who had been preemies found that even after factoring in maternal IQ, the kids who were breast-fed had more than an eight-point IQ advantage over their strictly formula-fed peers. Using MRI scans to measure their brains, the researchers found that in all the kids, but most clearly in boys, breast milk influenced the growth of white matter in the brain. The higher the percentage of breast milk in their diet as infants, the more white matter they had in their brains as teenagers, and the higher they scored on an intelligence test. Interestingly, the composition of human milk changes according to need, and preterm babies need more than mature ones do. It happens that Mother Nature packs more omega-3s into preterm milk.

Omega-3 fatty acids are not the only food for thought. Breast milk is also spiked with brain-boosting hormones and nerve growth factors. For instance, there's sialic acid—used in dendrites and synapses, the neural infrastructure of thought and memory. Human milk is thick with cholesterol, which forms myelin, part of the brain's white matter. Myelin wraps around neurons, protecting them, insulating them, and increasing the speed of signals, which loosely translates into intelligence. Myelin is an outgrowth of glial cells, which Einstein's brain was found to have in startling abundance. None of these ingredients are in formula.

I wouldn't be surprised if there's something magical in the milk. But there may also be an even simpler reason than maternal smarts behind the IQ boost. I'd wager that time and attention have something to do with it. Breast feeding is a real commitment: fifteen minutes or so on each breast, eight or more times a day, seven days a week. Any woman who finds it in herself to nurse for the first six months to a year of her child's life is probably the type to invest a lot of time, resources, and energy in rearing that child over the following eighteen years (not that formula-feeding moms don't).

The breast feeder finds herself stroking the feedee, singing and chatting in motherese. Could this physical and emotional stimulation have anything to do with the infant's intellectual development? Maybe. Maybe it tweaks how certain genes express themselves, as it does in mice that are licked and groomed frequently. Maybe the closeness translates into confidence, which adds a point or two on a standardized test.

All I know for sure is that my baby is voracious. Our breast-feeding sessions often go longer than expected, and most of the time she gets my undivided attention and affection. I swear the little suckler has learned to milk it.

Now that's smart.

Does Nursing Really Cause Sag?

In an episode of a reality TV show on teen pregnancy, the perky knocked-up cheerleader turns to her OB. "I'm not going to breast-feed. I want to save my breasts," she says. The doctor turns to her and in a motherly tone says something like: "Hon, it's not breast feeding that makes you sag. It's pregnancy."

According to the half-handful of studies on breast ptosis (droop), the doc's right. In one study, three out of four women experience lack of breast firmness after pregnancy, but there was no difference between women who breast-fed and those who didn't. In another, a team of plastic surgeons measured the degree of sag in nearly one hundred mothers, about half of whom breast-fed. Age, weight, greater number of pregnancies, larger prepregnancy bra size, and smoking were identified as significant independent risk factors for sag. Breast feeding was not.

I'm not convinced that breast feeding *never* contributes to sag—all that extra milk weight must do *something*—but it makes sense that pregnancy is the primary offender. Our breasts have undergone a major nine-month growth spurt. A lush forest of lobes and ducts developed under the influence of progesterone, prolactin, and lactogen. When we stop endowing our breasts with these hormones, which happens right after giving birth or after weaning, the lobules recede and the ducts dry up. The fat that had been displaced by all that mighty growth doesn't all go back to where it once was. It slumps.

Pregnancy does it, and the more pregnancies, the more ptosis. But age, gravity, and the loss of estrogen will cause it too. Get a good bra. Sag happens.

IS MILK A MOOD MANIPULATOR?

Not long ago, a babysitter fed my daughter some milk that I had pumped and stored in the freezer. The milk was several weeks old. I remember collecting it on a warm evening in early fall while chatting on the phone with a friend. We were howling about her recent induction into a society ladies' book club and tea. I had been gloating over my gorgeous offspring, then just a few weeks old. My nipples were like fountains that evening. My cups had overfloweth. The milk remained in the freezer until this afternoon when I went out, and the sitter defrosted it and fed it to our now three-month-old. She reportedly gulped down a bottleful and turned delightful. She gave the sitter a dreamy smile and drifted off to sleep in her arms. This wouldn't be so noteworthy if we had a baby who did not fuss before bedtime. "Wow, we drugged her," I said triumphantly.

I'm fascinated by the idea that there are "messages" in breast milk that manipulate an infant's temperament. Why not? In the past few years, this question has piqued the curiosity of a few scientists. It's a new area of research, but they've already found evidence that some hormones and other compounds in breast milk may be absorbed into the nursing baby's bloodstream, cross over to her brain, and influence her behavior. These hormones can carry messages about the environment, the mother's mood and status, and even the world at large—and how to act in it.

The simplest messages in milk may set a baby's clock, telling her if it's day or night. Morning milk is the more abundant. Evening milk is higher in fat, and, unlike day milk, it contains the hormone melatonin, which is released by the mother's pineal gland. As darkness falls, melatonin rises. Melatonin relaxes the baby, sending him into a flushed, slack-mouthed stupor. It is because of clock-setting melatonin (and day-night sensitive molecules called nucleotides) that breast-fed babies generally sleep better through the night than their formula-fed peers.

Funny enough, more melatonin also shows up in our breast milk when we laugh. Immunologist Hajime Kimata enrolled nearly ninety nursing mothers and their kids with allergies, and asked half of them to watch a Charlie Chaplin movie while the other half watched a video about the weather. The Japanese moms loved Chaplin, and they giggled throughout the viewing. Their breast milk was collected at set intervals throughout the day and the melatonin levels measured. It turned out that the Chaplin-loving mamas had significantly higher melatonin levels in their breast milk than the weather watchers. Babies fed "laughing milk" in turn had a significantly reduced allergic response to dust mites and other allergens. Kimata credited the melatonin in the milk with soothing the infants and programming their immune systems.

The message in melatonin-laced milk is mellowness. The breast milk I pumped that warm autumn eve, as I laughed with my friend on the phone, would have been dosed with melatonin. It said to my baby, "Sleep and heal. All is well in the universe, and it is time to rest."

A busy, stressed mom sends a very different signal to her baby. Her messenger is the hormone cortisol, and its message is to be alert and on edge. Cortisol ekes out of the adrenals into the bloodstream and into our breast milk. Like melatonin, it can cross over from the bloodstream into the baby's brain. The limbic areas of the brain that regulate behavior and emotion, especially the amygdala, are sensitive to cortisol.

According to a study by Laura Glynn and her colleagues at the University of California at Irvine, stressed-out moms who breast-fed their two-month-olds had babies who were more anxious and fearful of unfamiliar places and things, as rated by the mothers on a standardized test of infant behavior. The infants may startle more when they hear a sudden noise and are more likely to throw a tantrum when left alone. The stress is in the breast milk. When equally stressed-out high-cortisol moms fed their babies formula, those moms did not find any difference in their baby's behavior. The SOS isn't in the bottle; it's in the milk.

How horrible it seems for mother's milk to put a baby on edge—as if we're spiking it with espresso! But is it really so bad? From an evolutionary perspective, not really. The world we evolved in had steely-eyed predators, vengeful enemies, dodgy family dynamics, tragic weather. If you're a baby growing up in an unstable situation, it might be good for you to be a little more anxious when necessary. A steady drip of cortisol in the milk, and you're more alert, cautious, and fearful of new noises and strangers—behavior that might protect you from harm. Of course, the quantity of cortisol makes a big difference. If only moderate quantities of cortisol are in mama's milk, the hormone may help newborns regulate their nervous system and become more *resilient* to stress. Gender may play a role too: in monkeys, higher cortisol levels in breast milk are related to more confidence, though only in male infants, not females.

Another way a newborn can quickly learn about her world is by how much nutrition she gets in her mother's milk. In an intriguing study on rhesus monkeys at the University of California at Davis, infants adjusted their behavior and temperament depending on the quantity and the calorie and fat content in their mother's breast milk soon after birth. Regardless of social rank, mothers who weighed more and had previous pregnancies produced more milk, with more energy content, than mothers who weighed less and were less experienced. The infants who received the abundant and nutritious Grade A milk were more confident, curious, and active. These baby monkeys explored every nook and cranny of their pen. They were playful and sociable and handled separation and other stressful situations better than their peers did. The researchers speculated that babies adjust their behavioral patterns to conserve energy. The less nutritious the milk, the more nervous, inhibited, and sedentary the infant—presumably because the energy is better spent on growth and maintenance. These patterns were set in the first weeks of life and predicted the baby monkeys' behavior and temperament in the future.

Whether the quality of mother's milk influences human babies is unknown—and controversial. Could more caloric breast milk explain why second-borns are said to be more adventurous than firstborns? What could this mean on a population level, comparing breast milk quality in well-nourished versus poor countries? Maybe something, maybe nothing. Scientists acknowledge that genetic factors or the mother's temperament may explain some of the behavioral differences. Perhaps lighter, less experienced moms also have different levels of cortisol in their breast milk.

Of course, diet may also affect mood or milk or the mood in milk, thanks to the molecules from the foods that we pass into our breast milk. Broccoli, coffee, spices, citrus, nuts, and cow's milk may make some babies fussy. Caffeine may make a baby twitchier or more active. Alcohol in breast milk flavors it (even a glassful may show up on a baby's breath) and could disrupt his sleep patterns, making him more irritable.

It's fascinating, the idea that babies adjust their mood and temperament depending on the hormones and nutrition in the milk they drink. If I could tweak the recipe, I'd laugh every time I want my daughter to descend into a delicious melatonin-enhanced sleep. I'd add a drop of cortisol to steel her nerves, not shatter them. My milk would be rich and voluminous enough for her body to know she'll have abundant energy to explore. And her world would be the land of milk and honey.

Is Our Sweat Sexy?

It's one thing for my breast milk to influence my baby's moods, but to put strangers in the mood? Now that's power. Strange but true. A team of biologists at the Institute for Mind and Biology at the University of Chicago recruited nursing mothers and asked them to tuck pads into their nursing bras (over their nipples) and armpits to collect sweat and breast milk. The pads were collected for several months and given to childless young female volunteers to sniff every morning and evening. The sniffers kept a record of their sexual activity and erotic fantasies. They had no idea what they were inhaling.

By the second month of the study, the difference between the group smelling the breast-feeding mother's pads and a control group smelling a placebo was significant. Among the women exposed regularly to breast and nipple secretions, those with regular sex partners experienced a 24 percent increase in sexual desire as measured on a standard psychological test, and women without partners experienced a 17 percent increase in sexual fantasies. The hypersexual effect continued through the second half of the women's menstrual cycles, when libido normally declines.

The evolutionary purpose of the "aphrodisiac effect" is unknown, but it may be related to synchronized birthing in an ancestral setting. If breast-feeding odors subconsciously arouse other women in the mother's community, her female friends and relatives would be more likely to get pregnant and give birth around the same time. These mothers would support one another, increasing their children's chances of survival. In the past when food resources were unstable, breast-feeding odors may have signaled to other women that it was a safe time, calorically speaking, to get pregnant. See, I did it, you can too.

If breast-feeding odors push women toward maternity, perhaps they also push men toward paternity (more research is underway). Then everyone would feel like getting it on—except, perhaps, the new mom.

> I contain multitudes.
>   —Walt Whitman, "Song of Myself"

WHAT DO FETUSES LEAVE BEHIND?

My baby is now three months old. Officially, she's no longer a newborn. The leaves on the trees, so green and glorious the day of her birth, are now yellow and red. Slow down, stop! I look back at the giddy photos from the past summer—me in my postpartum pants and ponytail carrying my bundle under a dazzling magnolia tree, my husband with her in the crook of his arm. The baby—so tiny then; what happened?—was wearing newborn outfits that had been cleaned and caressed months before her birth, in wondrous expectation of her arrival, a coming that has already gone. Soon she'll be crawling and teething and walking and running off to college or whatever she wants to do with her life.

Is it any solace to us prenostalgic mothers that long after the umbilical cord has been cut, our babies will always be part of us, just as part of us will always be in our babies? I'm not talking about the emotional bonds between mother and child, which I can only hope will remain. I mean that some of our baby's cells may circulate in our bloodstreams for as long as we live. They may take residence in our spinal cord, skin, lungs, thyroid gland, liver, intestine, cervix, gallbladder, spleen, lymph nodes, and blood vessels. And, yes, our baby's cells can also live forever in our hearts and minds. Literally.

Here's what happens. During pregnancy, cells sneak across the placenta. The fetus's cells enter his mother, and the mother's cells enter the fetus. A baby's cells are detectable in his mother's bloodstream as soon as four weeks after conception, and a mother's cells are detectable in her fetus by week 13. Many of the fetus's cells that enter the mother are immune system cells, but some are stem cells. In the first trimester, one out of fifty thousand cells in your body would be from your baby-to-be (which is how some noninvasive prenatal tests check for genetic disorders). In the second and third trimesters, it's up to one fetal cell out of every thousand maternal cells. Around the time of the baby's birth, up to 6 percent of the DNA in your blood plasma is not your own; it comes from the fetus. Of this, some of the fetus's cells stick around for the long haul by creating their own lineages. Imagine colonies in the motherland. This is okay with us moms—our immune systems usually learn to tolerate our children's cells (as they tolerate ours). This is why skin and organ transplants between mother and child have a much higher success rate than between father and child.

Of course, we nosy mothers would like to know exactly what our children's cells are up to while they hang out in us. Are they just biding time in our bodies? Are they mother's little helpers? Or are they baby rebels, planning an insurgency?

When fetal cells are good, they are very, very good. They may reduce our risk of getting breast cancer, the disease that J. Lee Nelson and V. K. Gadi at the Hutchinson Cancer Research Center have studied. What they discovered is that fetal cells show up significantly more often in the breast tissue of women who don't have breast cancer than in women who do (43 versus 14 percent, respectively). The scientists propose that the baby's cells, which contain foreign DNA from her dad, stimulate the mother's immune system enough to continuously keep malignant cells in check. They prime us mothers to attack outsiders, not ourselves. This might explain why autoimmune diseases such as rheumatoid arthritis and multiple sclerosis improve during pregnancy and for some time afterward. The more fetal cells there are in a woman's body, the less active the disease.

Like our babies themselves, fetal cells have the potential to be anything. Some are stem cells. Stem cells have magical properties: they can morph into other types of cells (called differentiation), like liver, heart, or brain cells, and become part of those organs. Fetal stem cells migrate to injury sites—for instance, they've been found in diseased thyroid and liver tissue and have turned themselves into thyroid and liver cells. Are they on a special triage mission to repair the damage and repopulate the site? The current thinking is yes. Fetal cells may repair and rejuvenate us.

Then there's baby on the brain. This is the really startling stuff. Researchers working with mice have found evidence that cells from the fetus can cross a mother's brain-blood barrier and generate new neurons in the hippocampus, the memory area of the brain. If this happens in humans—and there's reason to believe it does—then it means, in a very real sense, that our babies integrate themselves into the circuitry of our minds. Could this help explain why new mothers grow new gray matter? Researchers thrill to the possibility of harnessing fetal cells to boost the brain, cure neurodegenerative diseases, and reverse the ravages of age. Maybe fetal cells are the real fountains of youth.

All is well when fetal cells are good, but when they are bad, they are horrid. They've shown up in cancers, and while they may be there to help, there's also a suspicion that they're not so innocent. While fetal cells may stimulate the immune system to be more vigilant, they may also make it overly aggressive. Our bodies may decide to attack the fetal cells within us, and in the crossfire healthy cells get bombarded. The fetal cells themselves may attack us, the little traitors. What sets off these battles is unknown, but in the fallout, we may suffer autoimmune diseases like scleroderma and lupus.

The maternal cells circulating in a baby's body are no more predictable. Nearly 1 in every 100 cells in a fetus comes from her mom. The population plummets to something like 1 in 100,000 after birth, but enough of a mother's special agents are still hiding out in her baby's tissues, and their ranks may be refreshed by refugees in breast milk that slip into the bloodstream. Maternal cells are busybodies. Some researchers think they train and shape the baby's immune system and even decrease the risk of allergies. They're healers too; there's evidence that maternal stem cells can morph into, for instance, insulin-producing cells that proliferate and repair damaged tissue in kids with juvenile diabetes. Like fetal cells in mothers, maternal cells in children may cause autoimmune problems.

When more than one person's cells mingle in one individual, the effect is known as microchimerism. The root of microchimerism is the "Chimera," an animal in Greek mythology. The Chimera is made up of the parts of multiple animals—and so, in a way, are we mothers. How many people have left their DNA in us? Any baby we've ever conceived, even ones we've miscarried unknowingly. Sons leave their Y chromosome genes in us. The cells of our older kids, flowing in our bloodstream, can be passed on to our younger kids. If we have an older sibling, that older sibling's cells may be in us. My mother's cells are in my body, and so are my daughter's cells, and half my daughter's cells come from her dad.

"Ha!" my husband chortles when I share this news. "That means I'm in you too." This is squirm-worthy. After all, some of that DNA may be in my brain.

But there's something beautiful about this too. Long post postpartum, we moms continue to carry our children in our bodies. My baby has become part of me, literally and figuratively, and she has changed me in ways I can't yet fathom. My sense of self has expanded (not just my waistline). The barriers have broken down; the lines are no longer fixed. Moms must be many in one.

That's the mother lode.

9

LESSONS FROM THE LAB

A Summary of Practical Tips

Use morning sickness as a diet guide.

Nausea and vomiting destroy the appetite, but we may have evolved to get morning sickness (actually around-the-clock sickness) in the first trimester to protect the embryo. Most toxins are tolerable in tiny quantities—in people with a mature and adequate immune system. But an embryo doesn't have one, so it needs to be protected. And the best way evolution could defend the embryo is by giving Mom nausea, vomiting, and food aversions. Foods least likely to trigger nausea—because they contain the fewest natural toxins—are simple carbs, grains, and starches. (Think of what kids like to eat.) Yet nutritious foods rich in B-complex vitamins and other vitamins and minerals (which may become more tolerable after first trimester) help the fetus develop and may even reverse epigenetic damage.

Hope the expectant father gets morning sickness too.

Nine of every ten expectant fathers are afflicted with at least one of their partner's symptoms: mood swings, nausea, fatigue, food cravings and aversions, and bouts of bloat. They often appear in the first trimester, often temporarily wane in the second trimester, and come back with a vengeance in the third trimester. Nearly half of these men gain thirty pounds or more of belly fat. It's not just sympathy; the hormone prolactin is behind the symptoms. Fathers-to-be have higher levels of the hormone prolactin, just as pregnant women do, which makes them feel sluggish and sensitive, and predisposed to gain weight easily. Compared to men whose prolactin levels are low or unchanged, men with escalating prolactin levels turn out to be more nurturing fathers. Among new dads, high-prolactin papas are found to be more attached to their newborns and more emotionally responsive than low-prolactin fathers. Expect the hormonal honeymoon to last four to seven weeks after the birth.

Prepare for a new persona.

Expect a transformation that goes beyond weight gain. You will not only look different, you'll also smell different, thanks to hormones and other chemical compounds that you and the fetus produce. Your favorite foods may become unappetizing, and your sense of smell keener. You'll bond better with other women and discover new talents in decoding emotions and recognizing faces. Faces you once thought were attractive may lose their allure. You're likely to be more attracted to people who look healthy or resemble your family. As worldly as you are, you may develop an unconscious bias against foreigners. As with food and odors, disgust applied to people peaks in the first trimester. You'll see the world through new eyes, and your focus will be on your safety and well-being.

Prepare for pregnesia.

You might forget details you'd otherwise remember—a condition known as "pregnesia." Many studies have found that pregnancy impairs both immediate and delayed recall. You may forget your future intentions, like taking your prenatal vitamins every day or what to pick up at the market on the way home. You might have more difficulty remembering street addresses, passages of poems, and the names of people standing in front of you. Some studies find that women pregnant with girls suffer from pregnesia more than those carrying boys.

Use your new mind-reading abilities.

Pregnant women are better at decoding other people's emotions—particularly anxiety and anger. According to the researchers, this may be an evolutionary adaptation to make moms-to-be more emotionally sensitive and vigilant toward signals of threat, aggression, and contagion. We're better at telling who might help us and who might hurt us. Babies with mind-reading mothers have an edge when it comes to getting what they want and need. Improved emotional intelligence may not only help us rally more support than we would otherwise, but also strengthen and expand our social networks.

Expect to hate the smell of your partner.

Prepregnancy, you probably liked the smell of your partner's body odor. It's a sign of compatibility on a biological level. It means he has immune system genes that complement your own. But in pregnancy (and when on the Pill), this preference appears to reverse. Hormones may be to blame. Maybe you're missing out on the raging, surging ovulation-related hormones that otherwise attract you to men who smell different from yourself. Or it could be that your nose is now under the influence of progesterone, which soars during pregnancy and on the Pill. Progesterone is associated with bonding and may draw us to kin. In our ancestral past, parents, siblings, cousins, uncles, and aunts may have been more helpful than mates when it comes to childbirth and raising a baby.

Embrace your nightmares.

Expect to have more vivid dreams, thanks to hormones and fractured sleep cycles. Also prepare yourself for more nightmares—and consider them a good thing. The evolutionary purpose of dreaming may be to help us resolve internal conflict and process new information, which is why we dream more when our lives are changing. Pregnant dreamers have a shorter labor than nondreamers—nearly an hour less on average. Among the dreamers, those who had vivid nightmares had significantly faster deliveries than those who had good dreams only. Women who had nightmares during pregnancy also had a significantly decreased chance of getting postpartum depression.

You may have unconsciously selected the gender of your baby.

But only unconsciously, only slightly, and never reliably. Male and female embryos thrive under different conditions. Very skinny women are more likely to have girls, and very bossy, dominant women appear to have boys more often. Married moms or moms living with their partners tend to have more sons, and single moms have more daughters. Women in rich nations have more sons on average, and women in poor nations have more daughters. Big and tall parents have more sons, and small and short parents have more daughters. Women who marry billionaires have more boys. Our hormones rise and fall. We age. Our weight goes up and down. Our body chemistry fluctuates yearly, monthly, daily, hourly. All this may influence the gender of the baby in subtle ways that are difficult to quantify and calculate.

Eat well, for what you eat (or don't eat) now could affect your baby's genes and even your grandchildren's genes.

Some foods contain chemicals (called methyl groups) that can attach themselves to a fetus's genes and control whether they turn on or off. Soy, leafy greens, meat, liver, shellfish, milk, leafy veggies, sunflower seeds, baker's yeast, egg yolk, liver, soy, beets, wheat, spinach, and shellfish are among the foods that have chemicals that do this. Adding methyl groups turns gene expression down (methylates) and removing them amps it up (demethylates), like turning the volume knob higher or lower, on or off. Egg yolk in particular contains choline, which is related to improved brain function. Fish contains omega-3s that may fuel brain growth. Excess fats may program your baby to eat more over a lifetime. A healthy diet may reverse some types of damage caused by toxins.

Don't OD on vitamins.

Excess vitamins in pregnancy may be harmful because they affect the ways the baby's genes behave. Overexuberant supplementation of folic acid and B vitamins during pregnancy is linked to higher rates of asthma. Some experts fear that excess multivitamin consumption in general, acting through epigenetic factors, may predispose children to obesity and metabolic syndrome. When pregnant rats on an otherwise normal diet are overfed multivitamins, their babies grew up hungrier and with little ability to control intake of food. Expectant moms exposed to a glut of vitamin E in supplements gave birth to babies with a five- to ninefold increased risk of congenital heart disease. Too much folic acid, vitamin B_{12}, soy, and other chemicals during sensitive periods in fetal development are linked with asthma and other complications. They work by altering the behavior of genes. Until we know the correct dosage and timing, a cure can also be a poison.

Feed your fetus fish.

The benefits of eating high-omega-3, sustainable, low-mercury, low-toxin fish such as wild salmon, sardines, and anchovies outweigh any risks of mercury poisoning. Many studies point to the brain benefits of at least a minimal amount of omega-3 fatty acids found in fish. In one Harvard study, kids who ate fish more than twice a week had improved performance on the cognitive tests regardless of their moms' mercury levels. Recent studies find that garlic, bran, and tea significantly reduce mercury absorption. Looked at one way, our brains even resemble fish flesh: the omega-3 fat profile in fish is closer to that of the human brain than any other food known. Neurons use these fatty acids to build their cell walls. One study found that omega-3s accounted for twice as much improvement in girls' test scores as in boys'.

Cultivate your baby's taste buds.

The foods you eat flavor your amniotic fluid. Sweets, spices, garlic, bitter veggies—whatever you taste, the baby tastes. Babies are hardwired to better detect and favor flavors they experienced in utero. Flavors they encounter in the womb may even cause structural changes in the brain that make them more sensitive to those flavors after birth. A sweet tooth and a salt habit can be programmed in fetuses. Even a taste for cigarettes and booze may be acquired in utero. These flavor preferences may stay with them for life.

Don't eat for two.

In pregnancy, the total weight gain for most women is between twenty-five and thirty-five pounds. This breaks down to about one hundred extra calories a day during the first trimester, the equivalent of a banana, and two hundred to three hundred extra calories a day in the second and third trimesters, the equivalent of two to three bananas. That's all. If we overeat in pregnancy, we risk not only gestational diabetes but preterm labor, hypertension, a C-section due to a monstrously large baby, or, almost as common, a small-for-gestational-age baby. Our prepregnancy weight may affect our baby's behavior in the future, for kids born of overweight moms have more risk of attention-deficit disorder. Even more disturbing, overeating may have an epigenetic effect on your baby. The fetus's genes that are related to appetite are overexpressed, which results in eating more, and the genes that control metabolism go haywire. Worse yet, overeating may affect your grandchildren's genes too.

Don't eat for just one either.

Aim to eat for 1.1 or 1.2. Eating too little (pregorexia) may have the same dangers for your baby as overeating—obesity, heart disease, and diabetes—because you're effectively programming him or her to hoard every calorie.

Your body odor may contain chemical signals that influence others subconsciously.

By the third trimester, you will sweat out (from your breasts and armpits) five chemical compounds. No one knows for sure why these chemicals show up in your sweat during pregnancy and disappear within six months of giving birth, but there's room for speculation. One theory is they help the newborn identify you, the mom, and helps him find your nipple so he can nurse soon after birth. Another theory is the subtle odor may subconsciously affect your partner. When a man kisses his pregnant wife, has sex with her, sniffs her sweat, or spends a great deal of time in her presence, he picks up on these chemical signals. (Your dog or cat may, too, which is why they're acting so protective.) Once inhaled, they'd be processed in his hypothalamus, the part of the brain that triggers the production of hormones. This may help explain why expectant fathers' hormone levels shift. Chemicals in your body odor may help you and your partner bond in the third trimester, just in time for the baby's birth.

Enjoy sex—pregnancy sex is a turn-on for most men and it can be good for the baby.

Don't worry; the baby will not know what you're up to. Some women worry that intercourse leads to miscarriage or early labor or otherwise may harm the fetus. *It does not in a pregnancy that is not high risk.* Women often feel less desirable during pregnancy, and needlessly so. In one study, 60 percent of men maintained the same level of perceived sexual desire for their pregnant partners, and 27 percent of men expressed increased desire. From an evolutionary perspective, sex during pregnancy is useful for moms and their babies. Simply put: fathers can be useful, and sex keeps them around. Sex is also an opportunity to practice Kegel exercises, the squeezing and relaxing of pelvic floor muscles, which speeds delivery, helps reduce complications, and reduces the risk of muscle laxity and incontinence after birth. Sex (even just oral sex) with the baby's father may help ward off miscarriage because proteins in his semen may help habituate your immune system and therefore not attack the fetus.

Tell the father the baby looks like him.

It's proven: men favor and invest more in babies that look like they do. Sometimes the baby really does favor Daddy; sometimes she doesn't. It's a roll of the dice. The more a kid looks like a man, the more a man likes the kid. One study found that as the percentage of a guy's features increased from 12.5 to 50 percent in a child's face, the more he was willing to adopt or spend time and money on him or her, and less likely to punish him or her. In one study, researchers found that a child's face had to have a minimum of 25 percent of a man's features for that man to favor that child over any other. That's the equivalent proportion of genes he'd share with nieces, nephews, half-siblings, and grandchildren.

Expose your fetus to sounds that may become touchstones.
By twenty-three weeks into the pregnancy, a fetus hears sounds outside the womb. Even if a one-year-old hasn't heard a song since she was
in the womb, she'll remember it. Babies exposed in utero to faster,
energetic music had the strongest preferences. Music also relaxes
fetuses directly and indirectly, by relaxing us. By the time the baby is
born, she knows the sound of her mother and her mother tongue, and
prefers them. She'll gravitate to the books you read her and the music
you played for her in utero. Listening to music and reading books won't
necessarily make her smarter, but familiar sounds will be a touchstone
for her after birth and may help calm her in her new alien world.

**Push yourself a little—in moderation, exercise and stress is
good for the fetus.**
Stress, both psychological and physical (exercise), can be good for the
developing fetus, but only in moderation. Several studies found that
moderate exercise and prenatal stress in mid- to late pregnancy are
associated with higher cognitive and motor scores. Compared to
women who had a somewhat stress-free nine months, moms who
reported moderate anxieties and pressures during the second and
third trimesters gave birth to babies with faster neural conduction for
three of the four tests of brain stem auditory evoked potential, an indicator of a more mature brain. Stress hormones, like fertilizer, may
result in new growth in the memory regions of the brain. Go ahead and
take on that keynote speaking opportunity; be competitive. Don't
worry about worrying.

**Avoid high stress for the sake of your child—and
grandchildren.**
Excessive stress can lower a baby's stress threshold, making him more
nervous and jittery after birth. High prenatal stress levels are also associated with a lower IQ, poorer health, and weaker maternal instinct
later in life. Worse yet, animal studies have found that stress-related
epigenetic changes made to a baby's genes in utero may be passed on

to subsequent generations. Happily, there's some evidence that the trauma of excessive prenatal stress may be reversed by an infant's secure attachment to her caregiver.

Eat a chocolate bar daily.

One study found that pregnant women who ate chocolate daily gave birth to babies who laughed easily, soothed readily, and responded well to novelty. Eating chocolate on a daily or regular basis during pregnancy may buffer a fetus from stress and improve his temperament after birth. The effect may be due to the psychoactive pleasure-producing stimulants in chocolate that relieve excess stress or compounds in it that prevent the breakdown of pleasure-producing chemicals in the brain. The sweet stuff also reduces the risk of preeclampsia, a condition that causes miscarriage. Dark chocolate is the healthiest and least fattening. If chocolate is your only caffeine source, it should be safe to gorge on about ten ounces of dark chocolate daily.

You can try to predict personality and intelligence in the womb.

Fidgety (or highly reactive) fetuses in third trimester may be fussier babies. Fetuses with slower and more variable heart rates at twenty-eight weeks gestation and after scored higher on mental and language development tests at age two-and-a-half years than did fetuses with faster and more stable heart rates. Spring-born babies consider themselves luckier in life, while fall- and winterborns, at least in youth, are found to be more risk-taking. The ratio of a person's fingers is established in the womb and related to prenatal testosterone levels. A ring finger that's longer than the index finger indicates high testosterone exposure and is linked to athleticism, aggression, and musical and mathematical ability.

Squat during childbirth (if you can).

Our ancestors almost certainly gave birth in an upright, squatting position. Squatting also expands the diameter of our skinny pelvises by 20 to 30 percent more than any other position and increases pressure on the abdomen, making it easier to push. Compared to the usual recumbent approach to childbirth, upright labor is 25 percent faster, decreases the risk of having to stretch and extend the vagina with an episiotomy, minimizes perineal tears and blood loss, reduces the need for an induction, and results in less severe pain.

Stimulating your breasts may trigger labor.

Not many home remedies are proven to trigger labor. Stimulating the nipples (by hand, pump, or mouth) may help, but only for women who are about to go into labor anyway. Nipple stimulation triggers oxytocin production, and oxytocin makes the uterus contract and expel. Several studies have found that this does indeed appear to make the uterus contract more strongly and the cervix ripen faster than it would otherwise. But it will likely work only if you're close to going into labor naturally.

Be prepared for the worst labor pains to happen after dark.

We may have evolved for labor to intensify at night because nighttime is when we're safest, calmest, and least distracted. Hormones are sensitive to circadian rhythms, and among those that peak at night are estriol and oxytocin. Estriol sets the scene for labor to happen. As it rises, progesterone, the custodian of a calm, contractionless uterus, backs down and the chaos begins. Meanwhile, like singles at last call, oxytocin receptors in the uterus are more concentrated and receptive in the wee hours of the morning. Our uteruses clench and cervixes dilate. Stress and distraction raise levels of adrenaline, which reduces oxytocin levels and thereby prevents labor.

Go skin-to-skin with your baby in the first hour after birth.

For the first hour after birth, the golden hour, press your newborn baby against your naked chest—skin-to-skin, heart-to-heart. In these first moments, the memory of your odor is emblazoned in the newborn's memory circuits. Minutes after birth, babies have a supersensitive sense of smell, and whatever odors they encounter—even unfamiliar ones—are imprinted in their memory. As the baby roots around and cries tears that may contain chemical signals, he triggers a cascade of subconscious hormonal responses in you. Two of those hormones are oxytocin and prolactin, which will make you produce milk, raise the temperature of your chest, and help you bond emotionally with your newborn.

Never bathe right after childbirth.

Newborns identify their moms by odor. Your breasts and nipples smell a bit like the only medium your baby has ever known: amniotic fluid. The chemical makeup of the broth that he's swallowed and marinated in for so long has some chemicals in common with nipple secretions, breast milk, and sweat. This smell is a source of comfort and perhaps the first way your baby knows you're you.

The more support you get during childbirth, the more likely you'll forget the pain later.

Five years after giving birth, about half of mothers remember childbirth as less painful than when they freshly delivered. These moms who forgot the pain were more likely to have had a positive experience than moms who remembered labor as equally or more painful than when they first rated it. Endorphins after delivery not only help us reduce or overlook pain in the moment, but also help us to forget the pain later. Moreover, we're likely to breed positive memories about the childbirth experience every time we think about it. These memories go forth and multiply.

Enjoy your new neurons.

Almost immediately after birth, your brain will likely undergo a massive makeover. In the brains of women who recently delivered, researchers found new gray matter in areas related to planning and execution, motivation and reward, perception and sensory integration, and motivation. Seeing, stroking, smelling the newborn, and even suckling him help trigger the growth of new neurons. The Mommy Brain is addicted to the newborn. In some ways, it's a less fearful and more aggressive brain than it was before your baby arrived. It's a mind that's more focused on the baby's safety and well-being. You may forget other things—shopping lists, names—because your brain may well specialize.

Manipulate your reward circuits.

Your own smiling baby does what no other baby can: light up your dopamine-driven reward-and-motivational circuits and areas associated with attachment. There is nothing more rewarding to most new mothers, especially in the postpartum months, than their baby's grin. Make a coffee table book of photos, or display a photo montage as your screensaver. It'll light up those reward circuits when you need a boost.

Spoil your newborn—for how she's treated now sets a foundation for the rest of her life.

Infancy is when babies form their impressions of how nurturing their world is. These impressions are psychological and biological, in that the amount of early care has an epigenetic effect. Animal studies have found that the more love and nurturing a baby gets early in life, the more receptors for stress hormones she gets, the higher her tolerance for stress, and the more motherly and nurturing she may be as a parent someday. Women who had nurturing mothers show more activity and gray matter volume in brain areas that can help them understand their own children's intentions and mental states. The most impressionable time for humans may be the first three years of life.

Become fluent in motherese.

Infants learn better when we speak to them in motherese—a slow, simple, repetitive speech spoken with exaggerated vowels in high pitch. Motherese helps attract babies' attention, they learn language faster, and they develop a larger vocabulary. The long, slow, stretched-out mouthing of words helps babies pick out vowel sounds and tell where one word begins and another ends.

Bathe the baby's brain and stomach in breast milk.

One million aggressive white blood cells swarm in every drop of mother's milk. Like red blood, it contains hormones, immune factors, vitamins, and minerals derived from your bloodstream. The immune forces of breast milk specialize in guarding the leaky gastrointestinal tract, where invaders creep in. Breast-fed babies are ten times less likely to be afflicted with infectious diarrhea because the walls of their gut are literally whitewashed with antibacterials. Breast-fed babies also have IQs that are nearly seven points higher on average than formula-fed babies, although the reason for the boost is unclear and may be due to maternal education. Nevertheless, breast milk is spiked with hormones, omega-3 fatty acids, and nerve growth factors that are absorbed into the baby's bloodstream and may influence brain development. The American Academy of Pediatrics recommends breast feeding for one year.

Be aware of the messages you send in your milk.
Hormones and other compounds in breast milk may be absorbed into the nursing baby's bloodstream, cross the brain-blood barrier, and influence the baby's behavior. These hormones may carry messages about the environment, your mood and status, and even the world at large—and how to act in it. In the evening, breast milk contains melatonin, a hormone that relaxes the baby and helps her sleep and heal. Melatonin levels are also higher in the breast milk of women who have been laughing. When we're busy and stressed, our breast milk contains cortisol. In moderation, the stress hormone in breast milk may be good for the baby's nervous system, making her more alert and even resilient under stress. In excess, cortisol may make the baby nervous, jumpy, and likely to throw a tantrum.

We are all Chimeras.
The Chimera is an animal made up of the parts of multiple animals—and so, it seems, are we. Some of our baby's cells have snuck across the placenta. There, they may circulate in our bloodstream or reside in our organs for the rest of our lives. Fetal cells may protect us from some diseases, including cancers, or they may be implicated in them. Our baby's cells may also reside in our brain, influencing our behavior—maybe our Mommy Brain—in ways science will someday explain.

SOURCES

1. STRETCH MARKS, SHRUNKEN BRAINS, AND A MOST SURPRISING SMELL— SCIENCE BEHIND THE SYMPTOMS

IS THERE A PURPOSE TO MORNING SICKNESS?

Boneva, R. S., Moore, C. A., Botto, L., Wong, L. Y., & Erickson, J. D. (1999). Nausea during pregnancy and congenital heart defects: A population-based case-control study. *Am J Epidemiol, 149*(8), 717–725.

Brown, J. E., Kahn, E. S., & Hartman, T. J. (1997). Profet, profits, and proof: Do nausea and vomiting of early pregnancy protect women from "harmful" vegetables? *Am J Obstet Gynecol, 176*(1 Pt 1), 179–181.

Cameron, E. L. (2007). Measures of human olfactory perception during pregnancy. *Chem Senses, 32*(8), 775–782.

Flaxman, S. M., & Sherman, P. W. (2008). Morning sickness: Adaptive cause or nonadaptive consequence of embryo viability? *Am Nat, 172*(1), 54–62.

Flaxman, S. M., & Sherman, P. W. (2000). Morning sickness: A mechanism for protecting mother and embryo. *Q Rev Biol, 75*(2), 113–148.

Young, S. (2011). *Craving earth: Understanding pica.* New York: Columbia University Press.

DO GIRLS MAKE US SICKER?

Yaron, Y., Lehavi, O., Orr-Urtreger, A., Gull, I., Lessing, J. B., Amit, A., et al. (2002). Maternal serum HCG is higher in the presence of a female fetus as early as week 3 post-fertilization. *Hum Reprod, 17*(2), 485–489.

WHY DOES DADDY-TO-BE STINK?

Gangestad, S. W., Garver-Apgar, C. E., Simpson, J. A., & Cousins, A. J. (2007). Changes in women's mate preferences across the ovulatory cycle. *J Pers Soc Psychol, 92*(1), 151–163.

Garver-Apgar, C. E., Gangestad, S. W., Thornhill, R., Miller, R. D., & Olp, J. J. (2006). Major histocompatibility complex alleles, sexual responsivity, and unfaithfulness in romantic couples. *Psychol Sci, 17*(10), 830–835.

Havlicek, J., & Roberts, S. C. (2009). MHC-correlated mate choice in humans: A review. *Psychoneuroendocrinology, 34*(4), 497–512.

Wedekind, C., & Penn, D. (2000). MHC genes, body odours, and odour preferences. *Nephrol Dial Transplant, 15*(9), 1269–1271.

Wedekind, C., Seebeck, T., Bettens, F., & Paepke, A. J. (1995). MHC-dependent mate preferences in humans. *Proc Biol Sci, 260*(1359), 245–249.

LOVE IN THE TIME OF GERMOPHOBIA

DeBruine, L. M., Jones, B. C., & Perrett, D. I. (2005). Women's attractiveness judgments of self-resembling faces change across the menstrual cycle. *Horm Behav, 47*(4), 379–383.

Jones, B. C., Perrett, D. I., Little, A. C., Boothroyd, L., Cornwell, R. E., Feinberg, D. R., et al. (2005). Menstrual cycle, pregnancy and oral contraceptive use alter attraction to apparent health in faces. *Proc Biol Sci, 272*(1561), 347–354.

Navarrete, C., Fessler, D., & Eng, S. (2007). Elevated ethnocentrism in the first trimester of pregnancy. *Evolution Hum Behav, 28*(1), 60–65.

DO WE SMELL PREGNANT?

Beauchamp, G., Curran, M., & Yamazaki, K. (2000). MHC-mediated fetal odourtypes expressed by pregnant females influence male associative behaviour .... *Anim Behav, 60*(3), 289–295.

Lucas, P. D., Donohoe, S. M., & Thody, A. J. (1982). The role of estrogen and progesterone in the control of preputial gland sex attractant odors in the female rat. *Physiol Behav, 28*(4), 601–607.

Vaglio, S. (2009). Chemical communication and mother-infant recognition. *Commun Integr Biol, 2*(3), 279–281.

WHAT'S CHANGING OUR LOOKS?

Aractingi, S., Berkane, N., Bertheau, P., Le Goue, C., Dausset, J., Uzan, S., et al. (1998). Fetal DNA in skin of polymorphic eruptions of pregnancy. *Lancet, 352*(9144), 1898–1901.

Barankin, B., Silver, S. G., & Carruthers, A. (2002). The skin in pregnancy. *J Cutan Med Surg, 6*(3), 236–240.

Buchanan, K., Fletcher, H. M., & Reid, M. (2010). Prevention of striae gravidarum with cocoa butter cream. *Int J Gynaecol Obstet, 108*(1), 65–68.

Nussbaum, R., & Benedetto, A. V. (2006). Cosmetic aspects of pregnancy. *Clin Dermatol, 24*(2), 133–141.

Osman, H., Usta, I. M., Rubeiz, N., Abu-Rustum, R., Charara, I., & Nassar, A. H. (2008). Cocoa butter lotion for prevention of striae gravidarum: A double-blind, randomised and placebo-controlled trial. *BJOG, 115*(9), 1138–1142.

Thornton, M. J. (2002). The biological actions of estrogens on skin. *Exp Dermatol, 11*(6), 487–502.

Youn, C. S., Kwon, O. S., Won, C. H., Hwang, E. J., Park, B. J., Eun, H. C., et al. (2003). Effect of pregnancy and menopause on facial wrinkling in women. *Acta Derm Venereol, 83*(6), 419–424.

IS THE FETUS TINKERING WITH OUR BRAIN?

Anderson, M. (2011). Recognition of novel faces after a single exposure is enhanced during pregnancy. *Ev Psych, 9*(1), 47–60.

de Groot, R. H., Hornstra, G., Roozendaal, N., & Jolles, J. (2003). Memory performance, but not information processing speed, may be reduced during early pregnancy. *J Clin Exp Neuropsychol, 25*(4), 482–488.

de Groot, R. H., Vuurman, E. F., Hornstra, G., & Jolles, J. (2006). Differences in cognitive performance during pregnancy and early motherhood. *Psychol Med, 36*(7), 1023–1032.

Kinsley, C. H., & Lambert, K. G. (2006). The maternal brain. *Sci Am, 294*(1), 72–79.

Macbeth, A. H., Gautreaux, C., & Luine, V. N. (2008). Pregnant rats show enhanced spatial memory, decreased anxiety, and altered levels of monoaminergic neurotransmitters. *Brain Res, 1241,* 136–147.

Oatridge, A., Holdcroft, A., Saeed, N., Hajnal, J. V., Puri, B. K., Fusi, L., et al. (2002). Change in brain size during and after pregnancy: Study in healthy women and women with preeclampsia. *Am J Neuroradiol, 23*(1), 19–26.

Pawluski, J. L., & Galea, L. A. (2007). Reproductive experience alters hippocampal neurogenesis during the postpartum period in the dam. *Neuroscience, 149*(1), 53–67.

Pearson, R. M., Lightman, S. L., & Evans, J. (2009). Emotional sensitivity for motherhood: Late pregnancy is associated with enhanced accuracy to encode emotional faces. *Horm Behav, 56*(5), 557–563.

WHY ARE OUR DREAMS MORE VIVID?

Brosh, A. (2003). Can dreams during pregnancy predict postpartum depression? *Dreaming, 13*(2), 66–72.

Garfield, P. (1988). *Women's bodies, women's dreams.* New York: Ballantine.

Mancuso, A., De Vivo, A., Fanara, G., Settineri, S., Giacobbe, A., & Pizzo, A. (2008). Emotional state and dreams in pregnant women. *Psychiatry Res, 160*(3), 380–386.

Maybruck, P. (1989). *Pregnancy and dreams: How to have a peaceful pregnancy by understanding your dreams, fantasies daydreams, and nightmares.* Los Angeles: Tarcher.

DO FETUSES DREAM?

Christos, G. A. (1995). Infant dreaming and fetal memory: A possible explanation of sudden infant death syndrome. *Med Hypotheses, 44*(4), 243–250.

Mirmiran, M. (1995). The function of fetal/neonatal rapid eye movement sleep. *Behav Brain Res, 69*(1–2), 13–22.

2. SKINNY CHICKS, BOSSY BROADS, AND A BASKETBALL IN THE BELLY— THE BIOLOGY OF BOY-OR-GIRL

COULD GIRLS (OR BOYS) RUN IN SOME FAMILIES?

Gellatly, C. (2009). Trends in population sex ratios may be explained by changes in the frequencies of polymorphic alleles of a sex ratio gene. *Evol Biol, 36*(2), 190–200.

Magnuson, A., Bodin, L., & Montgomery, S. M. (2007). Father's occupation and sex ratio of offspring. *Scand J Public Health, 35*(5), 454–459.

Weinberg, C. R., Baird, D. D., & Wilcox, A. J. (1995). The sex of the baby may be related to the length of the follicular phase in the conception cycle. *Hum Reprod, 10*(2), 304–307.

DO SKINNY CHICKS HAVE MORE DAUGHTERS?

Boklage, C. E. (2005). The epigenetic environment: Secondary sex ratio depends on differential survival in embryogenesis. *Hum Reprod, 20*(3), 583–587.

Bulik, C. M., Holle, A. V., Gendall, K., Lie, K. K., Hoffman, E., Mo, X., et al. (2008). Maternal eating disorders influence sex ratio at birth. *Acta Obstet Gynecol Scand, 87*(9), 979–981.

Cagnacci, A., Renzi, A., Arangino, S., Alessandrini, C., & Volpe, A. (2004). Influences of maternal weight on the secondary sex ratio of human offspring. *Hum Reprod, 19*(2), 442–444.

Cameron, E. Z., & Dalerum, F. (2009). A Trivers-Willard effect in contemporary humans: Male-biased sex ratios among billionaires. *PLoS ONE, 4*(1), e4195.

Coall, D. A., Meier, M., Hertwig, R., Wänke, M., & Höpflinger, F. (2009). Grandparental investment: The influence of reproductive timing and family size. *Am J Hum Biol, 21*, 455–463. doi: 10.1002/ajhb.20894.

Gibson, M. A., & Mace, R. (2003). Strong mothers bear more sons in rural Ethiopia. *Proc Biol Sci, 270* (Suppl 1), S108–109.

Kanazawa, S. (2005). Big and tall parents have more sons: Further generalizations of the Trivers-Willard hypothesis. *J Theor Biol, 235*(4), 583–590.

Mathews, F., Johnson, P. J., & Neil, A. (2008). You are what your mother eats: Evidence for maternal preconception diet influencing foetal sex in humans. *Proc Biol Sci, 275*(1643), 1661–1668.

Stein, A. D., Zybert, P. A., & Lumey, L. H. (2004). Acute undernutrition is not associated with excess of females at birth in humans: The Dutch hunger winter. *Proc Biol Sci, 271*(Suppl 4), S138–141.

Trivers, R. (2002). *Natural selection and social theory: Selected Papers of Robert Trivers.* New York: Oxford University Press.

DOUBLE X OR BUST?

Gray, R. H., Simpson, J. L., Bitto, A. C., Queenan, J. T., Li, C., Kambic, R. T., et al. (1998). Sex ratio associated with timing of insemination and length of the follicular phase in planned and unplanned pregnancies during use of natural family planning. *Hum Reprod, 13*(5), 1397–1400.

DO BOSSY BROADS HAVE MORE SONS?

Cameron, E. Z., & Dalerum, F. (2009). A Trivers-Willard effect in contemporary humans: Male-biased sex ratios among billionaires. *PLoS ONE, 4*(1), e4195.

Grant, V. (1998). *Maternal personality, evolution, and the sex ratio.* London: Routledge.

Grant, V. J. (2007). Could maternal testosterone levels govern mammalian sex ratio deviations? *J Theor Biol, 246*(4), 708–719.

Grant, V. J., Irwin, R. J., Standley, N. T., Shelling, A. N., & Chamley, L. W. (2008). Sex of bovine embryos may be related to mothers' preovulatory follicular testosterone. *Biol Reprod, 78*(5), 812–815.

Kanazawa, S., & Vandermassen, G. (2005). Engineers have more sons, nurses have more daughters: An evolutionary psychological extension of Baron-Cohen's extreme male brain theory of autism. *J Theor Biol, 233*(4), 589–599.

Manning, J. T., Martin, S., Trivers, R. L., & Soler, M. (2002). Second to 4th digit ratio and offspring sex ratio. *J Theor Biol, 217*(1), 93–95.

THE TRUTH ABOUT BASKETBALLS AND WATERMELONS

Perry, D. F., DiPietro, J., & Costigan, K. (1999). Are women carrying "basketballs" really having boys? Testing pregnancy folklore. *Birth, 26*(3), 172–177.

WHY WOULD GIRLS PREFER PINK?

Franklin, A., Bevis, L., Ling, Y., & Hurlbert, A. (2010). Biological components of colour preference in infancy. *Dev Sci, 13*(2), 346–354.

Frassanito, P., & Pettorini, B. (2008). Pink and blue: The color of gender. *Childs Nerv Syst, 24*(8), 881–882.

Hurlbert, A. C., & Ling, Y. (2007). Biological components of sex differences in color preference. *Curr Biol, 17*(16), R623–625.

Neitz, M., Kraft, T. W., & Neitz, J. (1998). Expression of L cone pigment gene subtypes in females. *Vision Res, 38*(21), 3221–3225.

3. FAT RATS, SALMON-AND-SOY SUPPERS, AND WHY GRANDMA'S DIET MATTERS— FOOD AND THE FETUS

CAN MOM'S DIET AFFECT BABY'S GENES?

Albright, C. D., Tsai, A. Y., Friedrich, C. B., Mar, M. H., & Zeisel, S. H. (1999). "Choline availability alters embryonic development of the hippocampus and septum in the rat." *Brain Res Dev Brain Res, 113*(1–2), 13–20.

Dolinoy, D. C., Weidman, J. R., Waterland, R. A., & Jirtle, R. L. (2006). Maternal genistein alters coat color and protects Avy mouse offspring from obesity by modifying the fetal epigenome. *Environ Health Perspect, 114*(4), 567–572.

Smedts, H. P., de Vries, J. H., Rakhshandehroo, M., Wildhagen, M. F., Verkleij-Hagoort, A. C., Steegers, E. A., et al. (2009). High maternal vitamin E intake by diet or supplements is associated with congenital heart defects in the offspring. *BJOG, 116*(3), 416–423.

Szeto, I. M., Das, P. J., Aziz, A., & Anderson, G. H. (2009). Multivitamin supplementation of Wistar rats during pregnancy accelerates the development of obesity in offspring fed an obesogenic diet. *Int J Obes (Lond), 33*(3), 364–372.

Warri, A., Saarinen, N. M., Makela, S., & Hilakivi-Clarke, L. (2008). The role of early life genistein exposures in modifying breast cancer risk. *Br J Cancer, 98*(9), 1485–1493.

Zeisel, S. H. (2009). Importance of methyl donors during reproduction. *Am J Clin Nutr, 89*(2), 673S—677S.

Zeisel, S. H. (2009). Epigenetic mechanisms for nutrition determinants of later health outcomes. *Am J Clin Nutr, 89*(suppl), 1488S–1493S.

THE BADNESS OF BPA
Dolinoy, D. C. (2007). Maternal nutrient supplementation counteracts bisphenol A—induced DNA hypomethylation in early development. *Proc Natl Acad Sci U S A, 104*(32), 13056–13061.

CAN WE EAT TOO MUCH FOR TWO?
Bayol, S. A., Simbi, B. H., Fowkes, R. C., & Stickland, N. C. (2010). A maternal "junk food" diet in pregnancy and lactation promotes nonalcoholic fatty liver disease in rat offspring. *Endocrinology, 151*(4), 1451–1461.
Cerf, M. E., & Louw, J. (2010). High fat programming induces glucose intolerance in weanling Wistar rats. *Horm Metab Res, 42*(5), 307–310.
Ludwig, D. S., & Currie, J. (2010). The association between pregnancy weight gain and birthweight: A within-family comparison *Lancet, 376*(9745), 984–990.
Mao, J., Zhang, X., Sieli, P. T., Falduto, M. T., Torres, K. E., & Rosenfeld, C. S. (2010). Contrasting effects of different maternal diets on sexually dimorphic gene expression in the murine placenta. *Proc Natl Acad Sci U S A, 107*(12), 5557–5572.
Ricart, W., Lopez, J., Mozas, J., Pericot, A., Sancho, M. A., Gonzalez, N., et al. (2009). Maternal glucose tolerance status influences the risk of macrosomia in male but not in female fetuses. *J Epidemiol Community Health, 63*(1), 64–68.
Rodriguez, A. (2010). Maternal pre-pregnancy obesity and risk for inattention and negative emotionality in children. *J Child Psychol Psychiatry, 51*(2), 134–143.

WHY GRANDMA'S (AND GRANDPA'S) DIET MATTERS
Kaati, G., Bygren, L. O., & Edvinsson, S. (2002). Cardiovascular and diabetes mortality determined by nutrition during parents' and grandparents' slow growth period. *Eur J Hum Genet, 10*(11), 682–688.
Lumey, L. H., Stein, A. D., Kahn, H. S., van der Pal-de Bruin, K. M., Blauw, G. J., Zybert, P. A., et al. (2007). Cohort profile: The Dutch Hunger Winter families study. *Int J Epidemiol, 36*(6), 1196–1204.
Pembrey, M. E., Bygren, L. O., Kaati, G., Edvinsson, S., Northstone, K., Sjostrom, M., et al. (2006). Sex-specific, male-line transgenerational responses in humans. *Eur J Hum Genet, 14*(2), 159–166.

DO PISCIVORES REALLY HAVE BRAINIER BABIES?
de Assis, S., Warri, A., Cruz, M. I., & Hilakivi-Clarke, L. (2010). Changes in mammary gland morphology and breast cancer risk in rats. *J Vis Exp, 21*(44), 2–4.
Broadhurst, C. L., Cunnane, S. C., & Crawford, M. A. (1998). Rift Valley lake fish and shellfish provided brain-specific nutrition for early *Homo. Br J Nutr, 79*(1), 3–21.
Broadhurst, C. L., Wang, Y., Crawford, M. A., Cunnane, S. C., Parkington, J. E., & Schmidt, W. F. (2002). Brain-specific lipids from marine, lacustrine, or terrestrial food resources: Potential impact on early African *Homo sapiens. Comp Biochem Physiol B Biochem Mol Biol, 131*(4), 653–673.
Furuhjelm, C., Warstedt, K., Larsson, J., Fredriksson, M., Bottcher, M. F., Falth-Magnusson, K., et al. (2009). Fish oil supplementation in pregnancy and lactation may decrease the risk of infant allergy. *Acta Paediatr, 98*(9), 1461–1467.
Helland, I. B., Smith, L., Blomen, B., Saarem, K., Saugstad, O. D., & Drevon, C. A. (2008). Effect of supplementing pregnant and lactating mothers with n-3 very-long-chain fatty acids on children's IQ and body mass index at 7 years of age. *Pediatrics, 122*(2), e472–479.
Hibbeln, J. R., & Davis, J. M. (2009). Considerations regarding neuropsychiatric nutritional requirements for intakes of omega-3 highly unsaturated fatty acids. *Prostaglandins Leukot Essent Fatty Acids, 81*(2–3), 179–186.
Hibbeln, J. R., Davis, J. M., Steer, C., Emmett, P., Rogers, I., Williams, C., et al. (2007). Maternal seafood consumption in pregnancy and neurodevelopmental outcomes in childhood (ALSPAC study): An observational cohort study. *Lancet, 369*(9561), 578–585.
Judge, M. P., Harel, O., & Lammi-Keefe, C. J. (2007). Maternal consumption of a

docosahexaenoic acid-containing functional food during pregnancy: Benefit for infant performance on problem-solving but not on recognition memory tasks at age 9 mo. *Am J Clin Nutr, 85*(6), 1572–1577.

Kannass, K. N., Colombo, J., & Carlson, S. E. (2009). Maternal DHA levels and toddler free-play attention. *Dev Neuropsychol, 34*(2), 159–174.

Lee, J. H., & Kang. H. S. (1999). Protective effects of garlic juice against embryotoxicity of methylmercuric chloride administered to pregnant Fischer 344 rats. *Yonsei Med J, 40*(5), 483–489.

Oken, E. (2009). Maternal and child obesity: The causal link. *Obstet Gynecol Clin North Am, 36*(2), 361–377, ix—x.

Shim, S. M., Feruzzi, M., et al. (2009) Impact of phytochemical-rich foods on bioaccessibility of mercury from fish. *Food Chemistry, 112*(2), 46–50.

BOOTY AND BRAINS

Lassek, W. D., & Gaulin, S. J. (2006). Changes in body fat distribution in relation to parity in American women: A covert form of maternal depletion. *Am J Phys Anthropol, 131*(2), 295–302.

WILL WHAT WE EAT NOW INFLUENCE BABY'S TASTES LATER?

Benassi, L., Accorsi, F., Marconi, L., & Benassi, G. (2004). Psychobiology of the amniotic environment. *Acta Biomed, 75* (Suppl 1), 18–22.

Browne, J. V. (2008). Chemosensory development in the fetus and newborn. *Newborn Infant Nursing Rev, 8*(4), 180–186.

Chotro, M. G., Arias, C., & Spear, N. E. (2009). Binge ethanol exposure in late gestation induces ethanol aversion in the dam but enhances ethanol intake in the offspring and affects their postnatal learning about ethanol. *Alcohol, 43*(6), 453–463.

Crystal, S., & Bernstein, I. L. (1995). Morning sickness, impact on offspring salt preference. *Appetite, 25,* 231–240.

Crystal, S. R., & Bernstein, I. L. (1998). Infant salt preference and mother's morning sickness. *Appetite, 30,* 297–307.

Fessler, D.M.T. (2003). An evolutionary explanation of the plasticity of salt preferences: Prophylaxis against sudden dehydration. *Medical Hypotheses, 61*(3), 412–415.

Malaga, I., Arguelles, J., Diaz, J. J., Perillan, C., Vijande, M., & Malaga, S. (2005). Maternal pregnancy vomiting and offspring salt taste sensitivity and blood pressure. *Pediatr Nephrol, 20,* 956–960.

Mennella, J. A., Jagnow, C. P., & Beauchamp, G. K. (2001). Prenatal and postnatal flavor learning by human infants. *Pediatrics, 107*(6), E88.

Schaal, B., Marlier, L., & Soussignan, R. (2000). Human foetuses learn odours from their pregnant mother's diet. *Chem Senses, 25*(6), 729–737.

IS A LITTLE TIPPLE REALLY SO TERRIBLE?

Ammon Avalos, Lyndsay, et al. (2009). Do multivitamin supplements modify the relationship between prenatal alcohol intake and miscarriage? *Am J Obst Gynecol, 201*(6), 563.e1–9.

Cohran, G., & Harpening, H. (2009). *The 10,000 year explosion.* New York: Basic Books.

Gemma, S., Vichi, S., & Testai, E. (2007). Metabolic and genetic factors contributing to alcohol-induced effects and fetal alcohol syndrome. *Neurosci Biobehav Rev, 31*(2), 221–229.

Green, R. F., & Stoler, J. M. (2007). "Alcohol dehydrogenase 1B genotype and fetal alcohol syndrome: A HuGE minireview." *Am J Obst Gynecol, 197*(1), 12–25.

Jacobson, S. W., et al. (2006). Protective effects of the alcohol dehydrogenase-ADH1B allele in children exposed to alcohol during pregnancy. *J Ped, 148*(1), 30–7.

Kelly, Y., Sacker, A., Gray, R., Kelly, J., Wolke, D., & Quigley, M. A. (2009). Light drinking in pregnancy, a risk for behavioural problems and cognitive deficits at 3 years of age? *Int J Epidemiol, 38*(1), 129–140.

O'Leary, C. M., Nassar, N., Kurinczuk, J. J., & Bower, C. (2009). The effect of maternal alcohol consumption on fetal growth and preterm birth. *BJOG, 116*(3), 390–400.

4. PUDGY HUBBIES, MAN BOOBS, AND A THEORY ABOUT THE DADDY GENE—
A DADDYOLOGY

WHY IS HUBBY GETTING SICK AND CHUBBY?

Berg, S. J., & Wynne-Edwards, K. E. (2001). Changes in testosterone, cortisol, and estradiol levels in men becoming fathers. *Mayo Clin Proc, 76*(6), 582–592.

Berg, S. J., & Wynne-Edwards, K. E. (2002). Salivary hormone concentrations in mothers and fathers becoming parents are not correlated. *Horm Behav, 42*(4), 424–436.

Dawson, W. (1930). *The custom of couvade.* Manchester: Manchester University Press.

Delahunty, K. M., McKay, D. W., Noseworthy, D. E., & Storey, A. E. (2007). Prolactin responses to infant cues in men and women: Effects of parental experience and recent infant contact. *Horm Behav, 51*(2), 213–220.

Klein, H. (1991). Couvade syndrome: Male counterpart to pregnancy. *Int J Psychiatry Med, 21*(1), 57–69.

Polo, M. (1903). *Ser Marco Polo the Venetian concerning the kingdoms and marvels of the East* (Vol. 12). New York: Scribner.

Storey, A. E., Walsh, C. J., Quinton, R. L., & Wynne-Edwards, K. E. (2000). Hormonal correlates of paternal responsiveness in new and expectant fathers. *Evol Hum Behav, 21*(2), 79–95.

DOES OUR SCENT AFFECT OUR PARTNER SUBCONSCIOUSLY?

Jacob, S., Spencer, N., & Bullivant, S. (2004). Effects of breastfeeding chemosignals on the human menstrual cycle. *Human Reproduction, 19*(2), 422–429.

Spencer, N. A., McClintock, M. K., Sellergren, S. A., Bullivant, S., Jacob, S., & Mennella, J. A. (2004). Social chemosignals from breastfeeding women increase sexual motivation. *Horm Behav, 46*(3), 362–370.

Stowers, L., & Marton, T. (2005). What is a pheromone? Mammalian pheromones reconsidered. *Neuron, 46*(5), 699–702.

Vaglio, S. (2009). Chemical communication and mother-infant recognition. *Commun Integr Biol, 2*(3), 279–281.

Vaglio, S., Minicozzi, P., Bonometti, E., & Mello, G. (2009). Volatile signals during pregnancy: A possible chemical basis for mother–infant recognition. *J Chem Ecol, 35*(1), 131–139.

CAN MEN BREAST-FEED?

Daly, M. (1979). Why don't male mammals lactate? *J Theor Biol, 78*(3), 325–345.

Diamond, J. (1997). *Why is sex fun?: The evolution of human sexuality.* New York: Basic Books.

Kunz, T. H., & Hosken, D. J. (2009). Male lactation: Why, why not and is it care? *Trends Ecol Evol, 24*(2), 80–85.

FATHERLY FACES

Roney, J. R., Hanson, K. N., Durante, K. M., & Maestripieri, D. (2006). Reading men's faces: Women's mate attractiveness judgments track men's testosterone and interest in infants. *Proc Biol Sci, 273*(1598), 2169–2175.

IS THERE A DADDY GENE?

Lim, M. M., Hammock, E. A., & Young, L. J. (2004). The role of vasopressin in the genetic and neural regulation of monogamy. *J Neuroendocrinol, 16*(4), 325–332.

Lim, M. M., & Young, L. J. (2004). Vasopressin-dependent neural circuits underlying pair bond formation in the monogamous prairie vole. *Neuroscience, 125*(1), 35–45.

Walum, H., Westberg, L., Henningsson, S., Neiderhiser, J. M., Reiss, D., Igl, W., et al. (2008). Genetic variation in the vasopressin receptor 1a gene (AVPR1A) associates with pair-bonding behavior in humans. *Proc Natl Acad Sci U S A, 105*(37), 14153–14156.

IS PREGNANCY NATURALLY A TURN-OFF?

Allende, I. (1999). *Aphrodite: A memoir of the senses.* New York: Harper.

Bartellas, E., Crane, J. M., Daley, M., Bennett, K. A., & Hutchens, D. (2000). Sexuality and sexual activity in pregnancy. *BJOG, 107*(8), 964–968.

Diamond, J. (1998). *Why is sex fun? The evolution of human sexuality.* New York: Basic Books.

Foux, R. (2008). *Pregnant sex.* N.p.: Erotic Review Books.

Stoppard, M. (1998). *Healthy pregnancy.* New York: DK Publishers.

von Sydow, K. (2002). Sexual enjoyment and orgasm postpartum: Sex differences and perceptual accuracy concerning partners' sexual experience. *J Psychosom Obstet Gynaecol, 23*(3), 147–155.

COULD SEX PREVENT MISCARRIAGE?

Davis, J. A., & Gallup Jr., G. G. (2006). Preeclampsia and other pregnancy related complications as an adaptive response to unfamiliar semen. In S. Platek & T. Shackelford (Eds.), *Female infidelity and paternal uncertainty: Evolutionary perspectives on male anti-cuckoldry tactics* (pp. 191–204). Cambridge: Cambridge University Press.

Kho, E. M., McCowan, L. M., North, R. A., Roberts, C. T., Chan, E., Black, M. A., et al. (2009). Duration of sexual relationship and its effect on preeclampsia and small for gestational age perinatal outcome. *J Reprod Immunol, 82*(1), 66–73.

Robertson, S. A., Bromfield, J. J., & Tremellen, K. P. (2003). Seminal "priming" for protection from pre-eclampsia: A unifying hypothesis. *J Reprod Immunol, 59*(2), 253–265.

Robillard, P. Y., Hulsey, T. C., Perianin, J., Janky, E., Miri, E. H., & Papiernik, E. (1994). Association of pregnancy-induced hypertension with duration of sexual cohabitation before conception. *Lancet, 344*(8928), 973–975.

5. MAMA'S BOYS, GREEDY FETUSES, AND WHY EVERYONE THINKS THE BABY LOOKS LIKE DAD— ON GENES AND BIASES

WHAT MAKES THE FETUS GREEDY?

Papper, Z., Jameson, N. M., et al. (2009). Ancient origin of placental expression in the growth hormone genes of anthropoid primates. Proc Natl Acad Sci U S A, 106(40), 17083–17088.

Haig, D. (2007). Placental growth hormone-related proteins and prolactin-related proteins. *Placenta, 29,* Suppl A, S36–41.

Wilkins, J., & Haig, D. (2003). What good is genomic imprinting: The function of parent-specific gene expression. *Nature Rev Genet, 4*(5), 359–368.

WHO CONTROLS BABY'S BRAIN?

Gregg, C. (2010). High-resolution analysis of parent-of-origin allelic expression in the mouse brain. *Science, 329,* 643–648.

Gregg, C., Zhang, J., Butler, J. E., Haig, D., & Dulac, C. (2010). Sex-specific parent-of-origin allelic expression in the mouse brain. *Science, 329*(5992), 682–685.

WHAT GENES DO BABIES INHERIT FROM MOM ONLY?

Kato, C., Umekage, T., Tochigi, M., Otowa, T., Hibino, H., Ohtani, T., et al. (2004). Mitochondrial DNA polymorphisms and extraversion. *Am J Med Genet B Neuropsychiatr Genet, 128B*(1), 76–79.

Shao, L., Martin, M. V., Watson, S. J., Schatzberg, A., Akil, H., Myers, R. M., et al. (2008). Mitochondrial involvement in psychiatric disorders. *Ann Med, 40*(4), 281–295.

HOW ARE ALL SONS MAMA'S BOYS?

Angier, N. (1999). *Woman: An intimate geography.* New York: Houghton Mifflin.

Zechner, U., Wilda, M., et al. (2001). A high density of X-linked genes for general cognitive ability: A run-away process shaping human evolution? *Trends in Genetics, 17*(12), 697–701.

DO MEN PREFER BABIES WHO RESEMBLE THEM?

Alvergne, A., Faurie, C. (2009). Father-offspring resemblance predicts paternal investment in humans. *Anim Behav, 76*(1), 61–69.

Burch, R., & Platek, S. M. (2006). The effect of perceived resemblance and the social mirror on kin selection. In S. M. Platek & T. Shackelford (Ed.), *Female infidelity and paternal uncertainty*. Cambridge: Cambridge University Press.

DeBruine, L. (2004). Resemblance to self increases the appeal of child faces to both men and women. *Evol Hum Behav, 25*(1), 142–154.

Heijkoop, M., & Dubas, J. (2009). Parent-child resemblance and kin investment: Physical resemblance or personality similarity? *Eur J Dev Psychol, 6*, 64–69.

Platek, S. M., Keenan, J. P., & Mohamed, F. B. (2005). Sex differences in the neural correlates of child facial resemblance: An event-related fMRI study. *Neuroimage, 25*(4), 1336–1344.

Platek, S. M., Raines, D. M., & Gallup, G. G. (2004). Reactions to children's faces: Males are more affected by resemblance than females are, and so are their brains. *Evol Hum Behavior, 25*(6), 394–405.

Volk, A. (2007). Parental investment and resemblance: Replications, refinements, and revisions. *Evol Psychol, 5*(1), 453–464.

WILL THE BABY REALLY LOOK MORE LIKE DAD?

Brédart, S., & French, R. (1999). Do babies resemble their fathers more than their mothers? A failure to replicate Christenfeld and Hill, *Evol Hum Behav, 2*(20), 129–135.

Bressan, P. (2002). Why babies look like their daddies: Paternity uncertainty and the evolution of self-deception in evaluating family resemblance. *Acta Ethologica, 4*, 113–118.

Christenfeld, N. J., & Hill, E. A. (1995). Whose baby are you? *Nature, 378*(6558), 669.

Daly, M., & Wilson, M. (1982). Whom are newborn babies said to resemble? *Ethol Sociobiology, 3*(2), 69–78.

Hayward, L., & Rohwer, S. (2004). Sex difference in attitudes toward paternity testing. *Evol Hum Behav, 25*(4), 242–248.

Heijkoop, M., & Dubas, J. (2009). Resemblance and kin investment. *Eur J Dev Psychol, 6*(1), 64–69.

DO GRANDPARENTS (UNCONSCIOUSLY) PLAY FAVORITES?

Clingempeel, W. G., Colyar, J. J., Brand, E., & Hetherington, E. M. (1992). Children's relationships with maternal grandparents: A longitudinal study of family structure and pubertal status effects. *Child Dev, 63*(6), 1404–1422.

Hawkes, K., O'Connell, J. F., Jones, N. G., Alvarez, H., & Charnov, E. L. (1998). Grandmothering, menopause, and the evolution of human life histories. *Proc Natl Acad Sci U S A, 95*(3), 1336–1339.

Hrdy, S. (2009). *Mothers and others: The evolutionary origins of mutual understanding*. Cambridge, MA: Belknap Press of Harvard University Press.

Kachel, A. F., Premo, L. S., & Hublin, J-J. (2010). Grandmothering and natural selection. *Proc R Soc B, 278*, 384–391.

Kuhle, B. X. (2007). An evolutionary perspective on the origin and ontogeny of menopause. *Maturitas, 57*(4), 329–337.

McBurney, D. (2002). Matrilateral biases in the investment of aunts and uncles. *Hum Nature, 13*(3), 391–395.

Pashos, A. (2000). Does paternal uncertainty explain discriminative grandparental solicitude? A cross-cultural study in Greece and Germany. *Evol Hum Behav, 21*(2), 97–109.

Sear, R. (2000). Maternal grandmothers improve nutritional status and survival of children in rural Gambia. *Proc R Soc B, 267*, 1641–1647.

6. FRAZZLED FETUSES, SNOOPING GENIUSES, AND WHY CHOCOLATE LOVERS HAVE SWEETER BABIES— PRENATAL PREDICTORS

WHAT DO FETUSES LEARN BY EAVESDROPPING?

Ando, Y., & Hattori, H. (1977). Effects of noise on sleep of babies. *JASA, 62,* 199–204.

Beauchemin, M., et al. (2010). Mother and stranger: An electrophysiological study of voice processing in newborns. *Cerebral Cortex* [Epub ahead of print]

DeCasper, A. J., & Fifer, W. P. (1980). Of human bonding: Newborns prefer their mothers' voices. *Science, 208*(4448), 1174–1176.

Hepper, P. G. (1988). Fetal "soap" addiction. *Lancet, 1*(8598), 1347–1348.

Kisilevsky, B. S., Hains, S. M., Brown, C. A., Lee, C. T., Cowperthwaite, B., Stutzman, S. S., et al. (2009). Fetal sensitivity to properties of maternal speech and language. *Infant Behav Dev, 32*(1), 59–71.

Lamont, A. (2003). Toddlers' musical preferences: Musical preference and musical memory in the early years. *Ann N Y Acad Sci, 999,* 518–519.

Lamont, A., & Dibben, N. (2001). Motivic structure and the perception of similarity. *Music Perception, 18,* 245–274.

Mampe, B., Friederici, A. D., Christophe, A., & Wermke, K. (2009). Newborns' cry melody is shaped by their native language. *Curr Biol, 19*(23), 1994–1997.

Pena, M., Maki, A., Kovacic, D., Dehaene-Lambertz, G., Koizumi, H., Bouquet, F., et al. (2003). Sounds and silence: An optical topography study of language recognition at birth. *Proc Natl Acad Sci U S A, 100*(20), 11702–11705.

WHAT CAN MOZART (OR ANY OTHER MUSIC) REALLY DO?

Carstens, C. B., Huskins, E., & Hounshell, G. W. (1995). Listening to Mozart may not enhance performance on the revised Minnesota Paper Form Board Test. *Psychol Rep, 77*(1), 111–114.

Chabris, C. F. (1999). Prelude or requiem for the "Mozart effect"? *Nature, 400*(6747), 826–827; author reply 827–828.

Clements, M. (1977). *Observations on certain aspects of neonatal behavior in response to auditory stimuli.* Paper presented at the 5th International Congress of Psychosomatic Obstetrics and Gynecology, Rome.

Fudin, R., & Lembessis, E. (2004). The Mozart effect: Questions about the seminal findings of Rauscher, Shaw, and colleagues. *Percept Mot Skills, 98*(2), 389–405.

Kim, H., Lee, M. H., Chang, H. K., Lee, T. H., Lee, H. H., Shin, M. C., et al. (2006). Influence of prenatal noise and music on the spatial memory and neurogenesis in the hippocampus of developing rats. *Brain Dev, 28*(2), 109–114.

Kisilevsky, B. S., Hains, S. M., Brown, C. A., Lee, C. T., Cowperthwaite, B., Stutzman, S. S., et al. (2009). Fetal sensitivity to properties of maternal speech and language. *Infant Behav Dev, 32*(1), 59–71.

Rauscher, F. H., Robinson, K. D., & Jens, J. J. (1998). Improved maze learning through early music exposure in rats. *Neurol Res, 20*(5), 427–432.

Rauscher, F. H., Shaw, G. L., & Ky, K. N. (1995). Listening to Mozart enhances spatial-temporal reasoning: Towards a neurophysiological basis. *Neurosci Lett, 185*(1), 44–47.

Rauscher, F. H., Shaw, G. L., Levine, L. J., Wright, E. L., Dennis, W. R., & Newcomb, R. L. (1997). Music training causes long-term enhancement of preschool children's spatial-temporal reasoning. *Neurol Res, 19*(1), 2–8.

Shetler, D. (1985). Prenatal music experiences. *Music Educators J, 71,* 26–27.

Steele, K. M., dalla Bella, S., Peretz, I., Dunlop, T., Dawe, L. A., Humphrey, G. K., et al. (1999). Prelude or requiem for the "Mozart effect"? *Nature, 400*(6747), 827–828.

Thompson, W. F., Schellenberg, E. G., & Husain, G. (2001). Arousal, mood, and the Mozart effect. *Psychol Sci, 12*(3), 248–251.

WILL EXERCISE STRENGTHEN BABY'S MIND?

Clapp, J. F., III. (1996). Morphometric and neurodevelopmental outcome at age five years of the offspring of women who continued to exercise regularly throughout pregnancy. *J Pediatr, 129*(6), 856–863.

Clapp, J. F. (2008). Exercise in pregnancy: A brief clinical review. *Fetal Matern Med, 2,* 89–101.

DeMaio, M., & Magann, E. F. (2009). Exercise and pregnancy. *J Am Acad Orthop Surg, 17*(8), 504–514.

Kim, H., Lee, S. H., Kim, S. S., Yoo, J. H., & Kim, C. J. (2007). The influence of maternal treadmill running during pregnancy on short-term memory and hippocampal cell survival in rat pups. *Int J Dev Neurosci, 25*(4), 243–249.

Parnpiansil, P., Jutapakdeegul, N., Chentanez, T., & Kotchabhakdi, N. (2003). Exercise during pregnancy increases hippocampal brain-derived neurotrophic factor mRNA expression and spatial learning in neonatal rat pup. *Neurosci Lett, 352*(1), 45–48.

Pinto, M. L., & Shetty, P. S. (1995). Influence of exercise-induced maternal stress on fetal outcome in Wistar rats: Inter-generational effects. *Br J Nutr, 73*(5), 645–653.

COULD FETUSES THRIVE ON STRESS?

Arai, J. & Feig, L. (2011). Long-lasting and transgenerational effects of an environmental enrichment on memory formation. *Brain Res Bul, 85*(1–2), 30–35.

Bergman, K., Sarkar, P., Glover, V., & O'Connor, T. G. (2010). Maternal prenatal cortisol and infant cognitive development: Moderation by infant-mother attachment. *Biol Psychiatry, 67*(11), 1026–1032.

Cannizzaro, C., Plescia, F., Martire, M., Gagliano, M., Cannizzaro, G., Mantia, G., et al. (2006). Single, intense prenatal stress decreases emotionality and enhances learning performance in the adolescent rat offspring: Interaction with a brief, daily maternal separation. *Behav Brain Res, 169*(1), 128–136.

Davis, E. P., & Sandman, C. A. (2010). The timing of prenatal exposure to maternal cortisol and psychosocial stress is associated with human infant cognitive development. *Child Dev, 81,* 131–148.

DiPietro, J. A. (2004). The role of prenatal maternal stress in child development. *Curr Dir Psychol Sci, 13,* 71–74.

DiPietro, J. A., Kivlighan, K. T., Costigan, K. A., & Laudenslager, M. L. (2009). Fetal motor activity and maternal cortisol. *Dev Psychobiol, 51*(6), 505–512.

DiPietro, J. A., Kivlighan, K. T., Costigan, K. A., Rubin, S. E., Shiffler, D. E., Henderson, J. L., & Pillion, J. P. (2010). Prenatal antecedents of newborn neurological maturation. *Child Dev, 81,* 115–130.

DiPietro, J. A., Novak, M. F. S. X., Costigan, K. A., Atella, L. D., & Reusing, S. P. (2006). Maternal psychological distress during pregnancy in relation to child development at age two. *Child Dev, 77,* 573–587

Fujioka, A., Fujioka, T., Ishida, Y., Maekawa, T., & Nakamura, S. (2006). Differential effects of prenatal stress on the morphological maturation of hippocampal neurons. *Neuroscience, 141*(2), 907–915.

Fujioka, T., Fujioka, A., Tan, N., Chowdhury, G. M., Mouri, H., Sakata, Y., et al. (2001). Mild prenatal stress enhances learning performance in the non-adopted rat offspring. *Neuroscience, 103*(2), 301–307.

Gutteling, B. M., de Weerth, C., Zandbelt, N., Mulder, E. J., Visser, G. H., & Buitelaar, J. K. (2006). Does maternal prenatal stress adversely affect the child's learning and memory at age six? *J Abnorm Child Psychol, 34*(6), 789–798.

Laplante, D. P., Brunet, A., Schmitz, N., Ciampi, A., & King, S. (2008). Project Ice Storm: Prenatal maternal stress affects cognitive and linguistic functioning in 5½-year-old children. *J Am Acad Child Adolesc Psychiatry, 47*(9), 1063–1072.

Lupien, S. J., McEwen, B. S., Gunnar, M. R., & Heim, C. (2009). Effects of stress throughout the lifespan on the brain, behaviour and cognition. *Nat Rev Neurosci, 10*(6), 434–445.

Räikkönen, K., Pesonen, A-K., Heinonen, K., et al. (2010). Maternal licorice consumption and detrimental cognitive and psychiatric outcomes in children. *Obstetrical & Gynecological Survey, 65*(2), 84–86.

Weinstock, M. (2008). The long-term behavioural consequences of prenatal stress. *Neurosci Biobehav Rev, 32*(6), 1073–1086.

HOW FETUSES CALM US
Entringer, S., Buss, C., Shirtcliff, E. A., Cammack, A. L., Yim, I. S., Chicz-Demet, A., et al. (2010). Attenuation of maternal psychophysiological stress responses and the maternal cortisol awakening response over the course of human pregnancy. *Stress, 13*(3), 258–268.

DO CHOCOLATE LOVERS HAVE SWEETER BABIES?
Connelly, M., Glavin, E., Lassner, A., and Paratore, J. (Executive Producers). (2009, September 30.). *The Ellen DeGeneres Show*. Burbank, CA: NBC Studios.

di Tomaso, E., Beltramo, M., & Piomelli, D. (1996). Brain cannabinoids in chocolate. *Nature, 382*(6593), 677–678.

Parker, G., Parker, I., & Brotchie, H. (2006). Mood state effects of chocolate. *J Affect Disord, 92*(2–3), 149–159.

Räikkönen K., Pesonen, A. K., Jarvenpaa, A. L., & Strandberg, T. E. (2004). Sweet babies: Chocolate consumption during pregnancy and infant temperament at six months. *Early Hum Dev, 76*(2), 139–145.

Trabucco, E., Acone, G., Marenna, A., Pierantoni, R., Cacciola, G., Chioccarelli, T., et al. (2009). Endocannabinoid system in first trimester placenta: Low FAAH and high CB1 expression characterize spontaneous miscarriage. *Placenta, 30*(6), 516–522.

Triche, E. W., Grosso, L. M., Belanger, K., Darefsky, A. S., Benowitz, N. L., & Bracken, M. B. (2008). Chocolate consumption in pregnancy and reduced likelihood of preeclampsia. *Epidemiology, 19*(3), 459–464.

DO FIDGETY FETUSES BECOME FEISTY KIDS?
Araki, M., Nishitani, S., Ushimaru, K., Masuzaki, H., Oishi, K., & Shinohara, K. (2010). Fetal response to induced maternal emotions. *J Physiol Sci, 60*(3), 213–220.

Davis, E. (2010). The timing of prenatal exposure to maternal cortisol and psychosocial stress is associated with human infant cognitive development. *Child Dev, 81*(1), 131–148.

Degani, S., Leibovitz, Z., Shapiro, I., & Ohel, G. (2009). Twins' temperament: Early prenatal sonographic assessment and postnatal correlation. *J Perinatology, 29*(5), 337–342.

DiPietro, J. A., Bornstein, M. H., Costigan, K. A., Pressman, E. K., Hahn, C. S., Painter, K., et al. (2002). What does fetal movement predict about behavior during the first two years of life? *Dev Psychobiol, 40*(4), 358–371.

DiPietro, J. A., & Costigan, K. A. (2003). Fetal response to induced maternal stress. *Early Hum Dev, 74,* 125–138.

DiPietro, J. A., Ghera, M. M., & Costigan, K. A. (2008). Prenatal origins of temperamental reactivity in early infancy. *Early Hum Dev, 84*(9), 569–575.

DiPietro, J., Hodgson, D., & Costigan, K. (1996). Fetal neurobehavioral development. *Child Dev, 67*(5), 2553–2567.

Gaultney, J. F., & Gingras, J. L. (2005). Fetal rate of behavioral inhibition and preference for novelty during infancy. *Early Hum Dev, 81*(4), 379–386.

WHAT DOES BABY'S BIRTH SEASON PREDICT?
Caci, H., Robert, P., & Boyer, P. (2004). Novelty seekers and impulsive subjects are low in morningness. *European Psychiatry*.

Cagnacci, A. (2005). Season of birth influences the timing of menopause. *Hum Repr, 20* (8), 2190–2193.

Chotai, J., & Adolfsson, R. (2002). Coverging evidence suggests that neurotransmitter turnover in human adults is associated with their season of birth. *Eur Arch Psychiatry Clin Neurosci, 252*(3), 130–134.

Chotai, J., Forsgren, T., Nilsson, L., & Adolfsson, R. (2001). Season of birth variations in the temperament and character inventory of personality in a general population. *Neuropsychobiology, 44*(1), 19–26.

Chotai, J., Joukamaa, M., & Taanila, A. (2009). Novelty seeking among adult women is lower for the winter borns compared to the summer borns: Replication in a large Finnish birth cohort. *ComprPsychiatry, 50*(6), 562–566.

Chotai, J., & Wiseman, R. (2005). Born lucky? The relationship between feeling lucky and month of birth. *Personality and Individual Differences, 39*(8), 1451–1460.

Dean, G., & Kelly, I. (2003). Is astrology relevant to consciousness and Psi? *J Consciousness Stud, 10,* 175–198.

Eisenberg, D., Campbell, B., & MacKillop, J. (2007). Season of birth and dopamine receptor gene associations with impulsivity, sensation seeking and reproductive behaviors. *PLos One, 11,* e1216.

Elter, K., Ay, E., Uyar, E., & Kavak, Z. N. (2004). Exposure to low outdoor temperature in the midtrimester is associated with low birth weight. *Aust N Z J Obstet Gynaecol, 44*(6), 553–557.

Huber, S. (2004). Effects of season of birth on reproduction in contemporary humans: Brief communication. *Human Reproduction, 19*(2), 445–447.

Joinson, C., & Nettle, D. (2005). Season of birth variation in sensation seeking in an adult population: Personality and individual differences. *J Hum Biol, 21*(2), 210–213.

Natale, V., Adan, A., & Chotai, J. (2007). Season of birth modulates mood seasonality in humans. *Psychiatry Res, 153*(2), 199–201.

Natale, V., Sansavini, A., & Trombini, E. (2005). Relationship between preterm birth and circadian typology in adolescence. *Neurosci Lett, 382*(1–2), 139–142.

WHAT CAN WE FORECAST FROM A FETUS'S FINGERS?

Blanchard, R. (2001). Fraternal birth order effect and the maternal immune hypothesis of male homosexuality. *Horm Behav, 40*(2), 105–114.

Kahn, H. S., Graff, M., Stein, A. D., Zybert, P. A., McKeague, I. W., & Lumey, L. H. (2008). A fingerprint characteristic associated with the early prenatal environment. *Am J Hum Biol, 20*(1), 59–65.

Lummaa, P., Pettay, J. E., & Russell, A. F. Male twins reduce fitness of female co-twins in humans. *Proc Natl Acad Sci U S A, 104*(26), 10915–10920.

Manning, J. T. (2002). *Digit ratio: A pointer to fertility, behavior, and health.* New Brunswick, NJ: Rutgers University Press.

Paul, S. N., Kato, B. S., Cherkas, L. F., Andrew, T., & Spector, T. D. (2006). Heritability of the second to fourth digit ratio (2d:4d): A twin study. *Twin Res Hum Genet,* 9(2), 215–219.

7. EVE'S LEGACY, NIPPLE POWER, AND THE GOLDEN HOUR— SOME SCIENCE OF THE MATERNITY WARD

IS THERE A PURPOSE TO PAINFUL BIRTH?

Maul, A. (2007). An evolutionary interpretation of the significance of physical pain experienced by human females: Defloration and childbirth pains. *Med Hypotheses, 69*(2), 403–409.

Ohel, I., Walfisch, A., Shitenberg, D., Sheiner, E., & Hallak, M. (2007). A rise in pain threshold during labor: a prospective clinical trial. *Pain, 132* (Suppl 1), S104–108.

Price, W. (2004). *Nutrition and physical degeneration.* Lemon Grove, CA: Price-Pottenger Nutrition Foundation.

Rosenberg, K., & Trevathan, W. (2002). Birth, obstetrics and human evolution. *BJOG, 109*(11), 1199–1206.

THE SCIENCE OF SQUAT

Bhardwaj, N. (1993). Squatting posture in traditional birth deliveries. *World Health Forum, 14*(4), 400–401.

Caldeyro-Barcia, R. (1979). The influence of maternal position on time of spontaneous rupture of the membranes, progress of labor, and fetal head compression. *Birth, 6*(1), 7–15.

Nasir, A., Korejo, R., & Noorani, K. J. (2007). Child birth in squatting position. *J Pak Med Assoc, 57*(1), 19–22.

Rosenberg, K. R., & Trevathan, W. R. (2001). The evolution of human birth. *Sci Am, 285*(5), 72–77.

DO WHITE MAMAS HAVE LONGER PREGNANCIES?

Falcao, V. (2003). Yes, natural gestation length has a strong genetic basis. *BMJ, 326,* 476.

Ma, S. (2008). Paternal race/ethnicity and birth outcomes. *Am J Public Health, 98*(12), 2285–2292.

Migone, A., Emanuel, I., Mueller, B., Daling, J., & Little, R. E. (1991). Gestational duration and birthweight in white, black and mixed-race babies. *Paediatr Perinat Epidemiol, 5*(4), 378–391.

Papiernik, E., Alexander, G. R., & Paneth, N. (1990). Racial differences in pregnancy duration and its implications for perinatal care. *Med Hypotheses, 33*(3), 181–186.

Patel, R. R., Steer, P., Doyle, P., Little, M. P., & Elliott, P. (2004). Does gestation vary by ethnic group? A London-based study of over 122,000 pregnancies with spontaneous onset of labour. *Int J Epidemiol, 33*(1), 107–113.

IS DADDY DELAYING US?

Haig, D. (2002). *Genomic imprinting and kinship.* New Brunswick, NJ: Rutgers University Press.

Lie, R. T., Wilcox, A. J., & Skjaerven, R. (2006). Maternal and paternal influences on length of pregnancy. *Obstet Gynecol, 107*(4), 880–885.

Migone, A., Emanuel, I., Mueller, B., Daling, J., & Little, R. E. (1991). Gestational duration and birthweight in white, black and mixed-race babies. *Paediatr Perinat Epidemiol, 5*(4), 378–391.

Olesen, A. W., Basso, O., & Olsen, J. (2003). Risk of recurrence of prolonged pregnancy. *BMJ, 326*(7387), 476.

CAN WE INDUCE OURSELVES?

Schaffir, J. (2002). Survey of folk beliefs about induction of labor. *Birth, 29*(1), 47–51.

Schaffir, J. (2006). Sexual intercourse at term and onset of labor. *Obstet Gynecol, 107*(6), 1310–1314.

Smith, C. A., & Crowther, C. A. (2004). Acupuncture for induction of labour. *Cochrane Database Syst Rev* (1), CD002962.

Tan, P. C., Yow, C. M., & Omar, S. Z. (2007). Effect of coital activity on onset of labor in women scheduled for labor induction: A randomized controlled trial. *Obstet Gynecol, 110*(4), 820–826.

Young, J. T., & Poppe, C. A. (1987). Breast pump stimulation to promote labor. *MCN Am J Matern Child Nurs, 12*(2), 124–126.

WHY DOES LABOR (OFTEN) STRIKE AT NIGHT?

Cassidy, T. P. (2007). *Birth: The surprising history of how we are born.* New York: Grove.

Navitsky, J., Greene, J. F., & Curry, S. L. (2000). The onset of human labor: Current theories. *Prim Care Update Ob Gyns, 7*(5), 197–199.

Ngwenya, S., & Lindow, S. W. (2004). Twenty-four hour rhythm in the timing of pre-labour spontaneous rupture of membranes at term. *Eur J Obstet Gynecol Reprod Biol, 112*(2), 151–153.

Odent, M. (2008). Birth and breastfeeding: *Rediscovering the needs of women during pregnancy and childbirth*. Sussex, UK: Rudolph Steiner Press.

Roizen, J., Luedke, C. E., Herzog, E. D., & Muglia, L. J. (2007). Oxytocin in the circadian timing of birth. *PLoS ONE*, 2(9), e922.

WHAT HAPPENS IN THE GOLDEN HOUR AFTER BIRTH?

Bramson, L., Lee, J. W., Moore, E., Montgomery, S., Neish, C., Bahjri, K., et al. (2010). Effect of early skin-to-skin mother-infant contact during the first 3 hours following birth on exclusive breastfeeding during the maternity hospital stay. *J Hum Lact, 26*(2), 130–137.

Bystrova, K., Ivanova, V., Edhborg, M., Matthiesen, A. S., Ransjo-Arvidson, A. B., Mukhamedrakhimov, R., et al. (2009). Early contact versus separation: Effects on mother-infant interaction one year later. *Birth, 36*(2), 97–109.

Chiu, S. H., Anderson, G. C., & Burkhammer, M. D. (2005). Newborn temperature during skin-to-skin breastfeeding in couples having breastfeeding difficulties. *Birth, 32*(2), 115–121.

Erlandsson, K., Dsilna, A., Fagerberg, I., & Christensson, K. (2007). Skin-to-skin care with the father after cesarean birth and its effect on newborn crying and prefeeding behavior. *Birth, 34*(2), 105–114.

Ferber, S. G., & Makhoul, I. R. (2004). The effect of skin-to-skin contact (kangaroo care) shortly after birth on the neurobehavioral responses of the term newborn: A randomized, controlled trial. *Pediatrics, 113*(4), 858–865.

Gelstein, S., et al. (2011). Human tears contain a chemosignal. *Science, 14*, 226–230.

Kaitz, M. (1992). Recognition of familiar individuals by touch. *Physiol Behav, 52*(3), 565–567.

Mizuno, K., Mizuno, N., Shinohara, T., & Noda, M. (2004). Mother-infant skin-to-skin contact after delivery results in early recognition of own mother's milk odour. *Acta Paediatr, 93*(12), 1640–1645.

Romantshik, O., Porter, R. H., Tillmann, V., & Varendi, H. (2007). Preliminary evidence of a sensitive period for olfactory learning by human newborns. *Acta Paediatr, 96*(3), 372–376.

Widstrom, A. M., Lilja, G., Aaltomaa-Michalias, P., Dahllof, A., Lintula, M., & Nissen, E. (2010). Newborn behavior to locate the breast when skin-to-skin: A possible method for enabling early self-regulation. *Acta Paediatr, 100*(1).

WHY IS BABY BORN BLUE-EYED?

Eiberg, H., et al. (2008). Blue eye color in humans may be caused by a perfectly associated founder mutation in a regulatory element located within the HERC2 gene inhibiting OCA2, *Hum Genet, 123*, 177–187.

ARE WE STRANGERS TO OUR NEWBORNS?

Cernoch, J. M., & Porter, R. H. (1985). Recognition of maternal axillary odors by infants. *Child Dev, 56*(6), 1593–1598.

Decasper, A. J., & Fifer, W. (2004). *Readings on the development of children* (4th ed.). New York: Macmillan.

Porter, R. H., & Winberg, J. (1999). Unique salience of maternal breast odors for newborn infants. *Neurosci Biobehav Rev, 23*(3), 439–449.

WHY MY BABY SMELLS SWEETER

Case, T. (2006). My baby doesn't smell as bad as yours. *Evol Hum Behav, 27*(5), 357–365.

Porter, R. H. (1998). Olfaction and human kin recognition. *Genetica, 104*(3), 259–263.

Porter, R. H., & Moore, J. D. (1981). Human kin recognition by olfactory cues. *Physiol Behav, 27*(3), 493–495.

DO WE REALLY FORGET THE PAIN?

Debiec, J., et al. (2006). Directly reactivated, but not indirectly reactivated, memories undergo reconsolidation in the amygdala. *Proc Natl Acad Sci, 103*(9), 3428–3433.

Ledoux, J. (1996). *The emotional brain*. New York: Simon & Schuster.

Norvell, K. T., Gaston-Johansson, F., & Fridh, G. (1987). Remembrance of labor pain: How valid are retrospective pain measurements? *Pain, 31*(1), 77–86.

Schiller, D., Monfils, M. H., Raio, C. M., Johnson, D. C., Ledoux, J. E., & Phelps, E. A. (2010). Preventing the return of fear in humans using reconsolidation update mechanisms. *Nature, 463*(7277), 49–53.

Waldenstrom, U., & Schytt, E. (2009). A longitudinal study of women's memory of labour pain—from 2 months to 5 years after the birth. *BJOG, 116*(4), 577–583.

8. MOMMY BRAIN, MOOD MILK, AND THE WEIRD HALF-LIFE OF CELLS THE BABY LEFT BEHIND— A POSTPARTUMOLOGY

DO MOMMIES HAVE BETTER BRAINS?

Bridges, E. S. (Ed.). (2008). *Neurobiology of the parental brain.* Orlando, FL: Academic Press.

Ellison, K. (2005). *The mommy brain: How motherhood makes us smarter.* New York: Basic Books.

Kim, P., Leckman, J. F., Mayes, L. C., Feldman, R., Wang, X., & Swain, J. E. (2010). The plasticity of human maternal brain: Longitudinal changes in brain anatomy during the early postpartum period. *Behav Neurosci, 124*(5), 695–700.

Kinsley, C. H., & Meyer, E. A. (2010). The construction of the maternal brain: Theoretical comment on Kim et al. (2010). *Behav Neurosci, 124*(5), 710–714.

Macbeth, A. H., & Luine, V. N. (2010). Changes in anxiety and cognition due to reproductive experience. *Neurosci Biobehav Rev, 34,* 452–467.

Narita, K., Takei, Y., Suda, M., Aoyama, Y., Uehara, T., Kosaka, H., et al. (2010). Relationship of parental bonding styles with gray matter volume of dorsolateral prefrontal cortex in young adults. *Prog Neuropsychopharmacol Biol Psychiatry, 34*(4), 624–631.

WHY MY BABY IS CUTEST

Aron, A., Fisher, H., Mashek, D. J., Strong, G., Li, H., & Brown, L. L. (2005). Reward, motivation, and emotion systems associated with early-stage intense romantic love. *J Neurophysiol, 94*(1), 327–337.

Bartels, A., & Zeki, S. (2004). The neural correlates of maternal and romantic love. *Neuroimage, 21*(3), 1155–1166.

Nitschke, J. B., Sarinopoulos, I., Mackiewicz, K. L., Schaefer, H. S., & Davidson, R. J. (2006). Functional neuroanatomy of aversion and its anticipation. *Neuroimage, 29*(1), 106–116.

Strathearn, L. (2008). What's in a smile? Maternal brain responses to infant facial cues. *Pediatrics, 122,* 40–51.

Swain, J. E. (2008). Baby stimuli and the parent brain: Functional neuroimaging of the neural substrates of parent-infant attachment. *Psychiatry (Edgmont), 5*(8), 28–36.

IS MATERNAL INSTINCT SHAPED IN INFANCY?

Ainsworth, M. S. (1965). *Further research into the adverse effects of maternal deprivation: Child care and growth of love.* New York: Penguin Books.

Ainsworth, M. S. (1970). Infant-mother attachment. *Amer Psychol, 34*(10), 932–937.

Kim, P., Leckman, J. (2010). Perceived quality of maternal care in childhood and structure and function of mothers' brains. *Dev Sci, 13*(4), 662–673.

O'Connell, A. N., & Russo, N. F. (1980). Models for achievement: Eminent women in psychology. *Psychology of Women Quarterly, 5,* 6–10.

Strathearn, L., Fonagy, P., Amico, J., & Montague, P. R. (2009). Adult attachment predicts maternal brain and oxytocin response to infant cues. *Neuropsychopharmacology, 34*(13), 2655–2666.

Szyf, M. F., Weaver, I., & Meaney M. (2007). Maternal care, the epigenome and phenotypic differences in behavior. *Repro Tox, 24*(1), 9–19.

Weaver, I., et al. (2004). Epigenetic programming by maternal behavior. *Nature, 7*(8), 847–854.

WHY DO WE SPEAK MOTHERESE?

Bettes, B. A. (1988). Maternal depression and motherese: Temporal and intonational features. *Child Dev, 59*(4), 1089–1096.

Falk, D. (2004). Prelinguistic evolution in early hominins: Whence motherese? *Behav Brain Sci, 27*(4), 491–503; discussion, 512–513.

Gordon, I., Zagoory-Sharon, O., Leckman, J. F., & Feldman, R. (2010). Oxytocin and the development of parenting in humans. *Biol Psychiatry, 68*(4), 377–382.

Hart, B., & Ridley, T. (1995). *Meaningful differences in the everyday experience of young American children.* Baltimore: Paul H. Brookes Pub. Co.

Kaplan, P. S., Bachorowski, J. A., Smoski, M. J., & Hudenko, W. J. (2002). Infants of depressed mothers, although competent learners, fail to learn in response to their own mothers' infant-directed speech. *Psychol Sci, 13*(3), 268–271.

Kemler Nelson, D. G., Hirsh-Pasek, K., Jusczyk, P. W., & Cassidy, K. W. (1989). How the prosodic cues in motherese might assist language learning. *J Child Lang, 16*(1), 55–68.

Schachner, A., & Hannon, E. E. (2011). Infant-directed speech drives social preferences in 5-month-old infants. *Dev Psychol., 47*(1), 19–25.

Trainor, L. J., & Desjardins, R. N. (2002). Pitch characteristics of infant-directed speech affect infants' ability to discriminate vowels. *Psychon Bull Rev, 9*(2), 335–340.

WHY ARE WE STRICKEN WITH COLIC AND DEPRESSION?

Doss, B. D., Rhoades, G. K., Stanley, S. M., & Markman, H. J. (2009). The effect of the transition to parenthood on relationship quality: An 8-year prospective study. *J Pers Soc Psychol, 96*(3), 601–619.

Gallup, G. G., Jr. (2010). Bottle feeding simulates child loss. *Med Hypotheses, 74,* 174–176.

Hagen, E. H. (1999). The functions of postpartum depression. *Evol Hum Behav, 20,* 325–359.

Hagen, E. H., & Barrett, H. C. (2007). Perinatal sadness among Shuar women: Support for an evolutionary theory of psychic pain. *Med Anthropol Q, 21*(1), 22–40.

Lummaa, V. (1998). Why cry? Adaptive significance of intensive crying in human infants. *Evol Hum Behav, 19,* 193–202.

Maxted, A. E., & Dicksten, S. (2005). Infant colic and maternal depression. *Infant Mental Health J, 26*(1), 56–68.

Miller, A. R., & Barr, R. G. (1991). Infantile colic: Is it a gut issue? *Pediatr Clin North Am, 38*(6), 1407–1423.

Newman, J. D. (2007). Neural circuits underlying crying and cry responding in mammals. *Behav Brain Res, 182*(2), 155–165.

Soltis, J. (2004). The signal functions of early infant crying. *Behav Brain Sci, 27*(4), 443–458; discussion 459–490.

WHY WE'RE LEFTIES WHEN WE CRADLE

Huggenberger, H. J., Suter, S. E., Reijnen, E., & Schachinger, H. (2009). Cradling side preference is associated with lateralized processing of baby facial expressions in females. *Brain Cogn, 70*(1), 67–72.

Manning, J. T., & Denman, J. (1994). Lateral cradling preferences in humans (*Homo sapiens*): Similarities within families. *J Comp Psychol, 108*(3), 262–265.

Reissland, N., Hopkins, B., Helms, P., & Williams, B. (2009). Maternal stress and depression and the lateralisation of infant cradling. *J Child Psychol Psychiatry, 50*(3), 263–269.

Suter, S. E., Huggenberger, H. J., Richter, S., Blumenthal, T. D., & Schachinger, H. (2009). Left side cradling of an appetitive doll is associated with higher heart rate variability and attenuated startle in nulliparous females. *Int J Psychophysiol, 74*(1), 53–57.

WHAT'S LIVING IN OUR MILK (AND WHY)?

Mossberg, A. K., et al. (2010). HAMLET interacts with lipid membranes and perturbs their structure and integrity. *PLoS ONE, 5* (2), e9384.

Savino, F., Liguori, S. A., Fissore, M. F., & Oggero, R. (2009). Breast milk hormones and their protective effect on obesity. *Int J Pediatr Endocrinol* (2009), 327505.

Schack-Nielsen, L., & Michaelsen, K. F. (2007). The effects of breastfeeding I: Effects on the immune system and the central nervous system. *Ugeskr Laeger, 169*(11), 985–989.

Sheard, N. F., & Walker, W. A. (1988). The role of breast milk in the development of the gastrointestinal tract. *Nutr Rev, 46*(1), 1–8.

Walker, A. (2010). Breast milk as the gold standard for protective nutrients. *J Pediatr, 156*(2 Suppl), S3–7.

ARE BREAST-FED BABIES REALLY BRAINIER?

Bartels, M., van Beijsterveldt, C. E., & Boomsma, D. I. (2009). Breastfeeding, maternal education and cognitive function: A prospective study in twins. *Behav Genet, 39*(6), 616–622.

Caspi, A. (2007). Moderation of breastfeeding effects on the IQ by genetic variation in fatty acid metabolism. *Proc Natl Acad Sci U S A, 104*(47), 18860–18865.

Isaacs, E. B., Fischl, B. R., Quinn, B. T., Chong, W. K., Gadian, D. G., & Lucas, A. (2009). Impact of breast milk on IQ, brain size and white matter development. *Pediatr Res. Pediatrics, 109*, 1044–1053.

Jain, A., Concato, J., & Leventhal, J. M. (2002). How good is the evidence linking breastfeeding and intelligence? *Pediatrics, 109*(6), 1044–1053.

Jensen, C. L., & Lapillonne, A. (2009). Docosahexaenoic acid and lactation. *Prostaglandins Leukot Essent Fatty Acids, 81*(2–3), 175–178.

Jensen, C. L., Voigt, R. G., Llorente, A. M., Peters, S. U., Prager, T. C., Zou, Y. L., et al. (2010). Effects of early maternal docosahexaenoic acid intake on neuropsychological status and visual acuity at five years of age of breast-fed term infants. *J Pediatr, 157*(6), 900–905.

Lundberg, G. D. (2008). Does breast-feeding improve child cognitive development? *Medscape J Med, 10*(8), 197.

Steer, C. (2010). FADS2 Polymorphisms modify the effect of breastfeeding on child IQ. *PLoS Biol, 5*(7), e11570.

DOES NURSING REALLY CAUSE SAG?

Geddes, D. T. (2007). Inside the lactating breast: The latest anatomy research. *J Midwifery Womens Health, 52*(6), 556–563.

Pisacane, A. (2007). Breastfeeding and perceived changes in the appearance of breasts. *Acta Paediatr, 93*(10), 1346–1348.

Rinker, B., Veneracion, M., & Walsh, C. P. (2010). Breast ptosis: Causes and cure. *Ann Plast Surg, 64*(5), 579–584.

IS MILK A MOOD MANIPULATOR?

Cao, Y., Rao, S. D., Phillips, T. M., Umbach, D. M., Bernbaum, J. C., Archer, J. I., et al. (2009). Are breast-fed infants more resilient? Feeding method and cortisol in infants. *J Pediatr, 154*(3), 452–454.

Cubero, J., Valero, V., Sanchez, J., Rivero, M., Parvez, H., Rodriguez, A. B., et al. (2005). The circadian rhythm of tryptophan in breast milk affects the rhythms of 6-sulfatoxymelatonin and sleep in newborn. *Neuro Endocrinol Lett, 26*(6), 657–661.

Glynn, L., Davis, L., et al. (2007). Postnatal maternal cortisol levels predict temperament in healthy breastfed infants. *Early Hum Dev, 83*, 675–681.

Hinde, K., & Capitanio, J. P. (2010). Lactational programming? Mother's milk energy predicts infant behavior and temperament in rhesus macaques (*Macaca mulatta*). *Am J Primatol, 72*(6), 522–529.

Kimata, H. (2007). Laughter elevates the levels of breast-milk melatonin. *J Psychosom Res, 62*(6), 699–702.

Peaker, M., & Neville, M. C. (1991). Hormones in milk: Chemical signals to the offspring? *J Endocrinol, 131*(1), 1–3.

Sanchez, C. L., Cubero, J., Sanchez, J., Chanclon, B., Rivero, M., Rodriguez, A. B., et al. (2009). The possible role of human milk nucleotides as sleep inducers. *Nutr Neurosci, 12*(1), 2–8.

Sullivan, E. (2011). Cortisol concentrations in the milk of rhesus monkey mothers are associated with confident temperament in sons, but not in daughters. *Developmental psychobiology, 53*(1), 96–104.

IS OUR SWEAT SEXY?

Spencer, N. A., McClintock, M. K., Sellergren, S. A., Bullivant, S., Jacob, S., & Mennella, J. A. (2004). Social chemosignals from breastfeeding women increase sexual motivation. *Horm Behav, 46*(3), 362–370.

WHAT DO FETUSES LEAVE BEHIND?

Fugazzola, L., Cirello, V., & Beck-Peccoz, P. (2010). Fetal cell microchimerism in human cancers. *Cancer Lett, 287*(2), 136–141.

Gadi, V. K. (2009). Fetal microchimerism and cancer. *Cancer Lett, 276*(1), 8–13.

Gadi, V. K. (2010). Fetal microchimerism in breast from women with and without breast cancer. *Breast Cancer Res Treat, 121*(1), 241–244.

Gadi, V. K., & Nelson, J. L. (2007). Fetal microchimerism in women with breast cancer. *Cancer Res, 1*(67)(19), 9035–9038.

Gammill, H. S., & Nelson, J. L. (2010). Naturally acquired microchimerism. *Int J Dev Biol, 54*(2–3), 531–543.

Lee, E. (2010). Fetal cell microchimerism: Natural-born killers or healers? *Mol Hum Reprod, 16*(11), 869–878.

Lissauer, D., Piper, K. P., Moss, P. A., & Kilby, M. D. (2007). Persistence of fetal cells in the mother: friend or foe? *BJOG, 114*(11), 1321–1325.

Lissauer, D. M., Piper, K. P., Moss, P. A., & Kilby, M. D. (2009). Fetal microchimerism: the cellular and immunological legacy of pregnancy. *Expert Rev Mol Med, 11*, e33.

Maurel, M. C., & Kanellopoulos-Langevin, C. (2008). Heredity: Venturing beyond genetics. *Biol Reprod, 79*(1), 2–8.

Zeng, X. X., Tan, K. H., Yeo, A., Sasajala, P., Tan, X., Xiao, Z. C., et al. (2010). Pregnancy-associated progenitor cells differentiate and mature into neurons in the maternal brain. *Stem Cells Dev, 19*(12), 1819–1830.

ACKNOWLEDGMENTS

I am grateful to all the researchers whose work on the science of pregnancy and parenthood inspired this book. I'd like to especially thank those who took the time to answer my questions, including the following people: Pilyoung Kim, Gary Beauchamp, Jayanti Chotai, Valerie Grant, Stefano Vaglio, Jaroslav Flegr, Christine Garver-Apgar, Marco Del Giudice, Bruno Laeng, Corry Gellatly, Suma Jacob, Ben Jones, Lisa DeBruine, and Gordon Gallup, Jr. Of course, any omissions or errors are my own.

Special thanks go to my editor at Free Press, Hilary Redmon, for her enthusiasm and support for this book from conception to delivery. I'm grateful to Sydney Tanigawa for her thoughtful edits and her kindness. My husband, Peter, is to be thanked for his love, comfort, and humor throughout the pregnancy and the creation of this book, not to mention his role in bringing our daughter into the world. Props go to the friends (and strangers) who shared their pregnancy experiences, questions, opinions, and advice. I'd like to especially thank my parents for use of their getaway spot in Vermont. My mom deserves my gratitude for watching the baby in those early weeks when I decided that I needed time out to write. Now I know what she went through when I allegedly had colic. She calls it karma.

INDEX

ABOUT THE AUTHOR

Jena Pincott has a background in biology and worked on the production of science documentaries for PBS. A former senior editor at Random House, she is the author of *Do Gentlemen Really Prefer Blondes: Bodies, Behavior, and Brains—The Science Behind Sex, Love, and Attraction,* which is being translated into more than fifteen languages; and *Success: Advice for Achieving Your Goals from Remarkably Accomplished People.* She lives in New York City with her husband and baby daughter, the inspiration for this book. Follow her blog on the science of love, sex, and babies at www.jenapincott.com.